Environmental Contamination Remediation and Management

There are many global environmental issues that are directly related to varying levels of contamination from both inorganic and organic contaminants. These affect the quality of drinking water, food, soil, aquatic ecosystems, urban systems, agricultural systems and natural habitats. This has led to the development of assessment methods and remediation strategies to identify, reduce, remove or contain contaminant loadings from these systems using various natural or engineered technologies. In most cases, these strategies utilize interdisciplinary approaches that rely on chemistry, ecology, toxicology, hydrology, modeling and engineering.

This book series provides an outlet to summarize environmental contamination related topics that provide a path forward in understanding the current state and mitigation, both regionally and globally.

Topic areas may include, but are not limited to, Environmental Fate and Effects, Environmental Effects Monitoring, Water Re-use, Waste Management, Food Safety, Ecological Restoration, Remediation of Contaminated Sites, Analytical Methodology, and Climate Change.

T. Vamsi Nagaraju • G. Sri Bala
Editors

Delta Ecosystems

From Contamination to Conservation

Springer

Editors
T. Vamsi Nagaraju
Department of Civil Engineering
SRKR Engineering College
Bhimavaram, Andhra Pradesh, India

G. Sri Bala
Department of Civil Engineering
SRKR Engineering College
Bhimavaram, Andhra Pradesh, India

ISSN 2522-5847 ISSN 2522-5855 (electronic)
Environmental Contamination Remediation and Management
ISBN 978-3-031-97919-4 ISBN 978-3-031-97920-0 (eBook)
https://doi.org/10.1007/978-3-031-97920-0

This Springer imprint is published by the registered company Springer Nature Switzerland AG
The registered company address is: Gewerbestrasse 11, 6330 Cham, Switzerland

Contents

Chapter 1
Delta Ecosystems: Water Quality and Environment

T. Vamsi Nagaraju and G. Sri Bala

Abstract Delta regions are dynamic ecosystems where rivers, drains, and canal discharges bring water into coastal areas. Water quality is an important metric that describes how desirable water is for drinking or domestic requirements. Many physical, chemical, and biological parameters are used to assess water quality, such as pH, total dissolved solids, electrical conductivity, dissolved oxygen, salinity, and nitrogenous compounds. Freshwater and marine waters are frequently mixed together in the estuaries, which is more concern. The water quality dynamics of delta ecosystems and their environmental effects are discussed in this chapter, with particular attention on the Mekong Delta in Vietnam and the Godavari Delta in India. Additionally, a future direction of water quality study in delta regions is indicated to significantly improve water quality worldwide, taking into account all the current information gaps relevant to the latest strategies.

Keywords Delta region · Inland aquaculture · Ecosystem · Water quality

1.1 Introduction

Water is an essential resource for human survival, economic development, and ecological balance. It is essential to agriculture, industry, and biodiversity and provides basic existence. However, water bodies, especially surface water, are under unprecedented stress due to human activities like industrialization, urbanization, and intensive agricultural practices (Mukherjee et al. 2020). Globally, the effects of these activities have been shown to include ecological deterioration, water pollution, and

T. V. Nagaraju (✉) · G. S. Bala
Department of Civil Engineering, SRKR Engineering College, Bhimavaram, Andhra Pradesh, India

Centre for Sustainable Inland Aquaculture, SRKR Engineering College, Bhimavaram, India
e-mail: tvnraju@srkrec.edu.in

T. V. Nagaraju, G. S. Bala (eds.), *Delta Ecosystems*, Environmental Contamination Remediation and Management, https://doi.org/10.1007/978-3-031-97920-0_1

depletion (Ogidi and Akpan 2022). The severity of the problem is demonstrated by the microbial pathogen and heavy metal contamination seen in the Sisa River near Kumasi, Ghana, and the degradation of Central Ethiopian rivers over the last 60 years (Takyi et al. 2022). Subsequently, low water quality is associated with high mortality rates and waterborne illnesses, and these trends are a worldwide possibility and have important public health complications (Takyi et al. 2022). Poor water quality, for instance, contributes to mortality rates of up to 8.5% in Southeast Asia (Lim and Nissapatorn 2017). Due to its accessibility and availability, surface water is frequently used for water treatment but is severely degraded (Akhtar et al. 2021). This degradation threatens ecosystems, human health, and water-dependent economic activities. Understanding water quality parameters is necessary to address these issues, especially in ecologically significant and sensitive areas like deltas (Giri 2021).

Delta habitats, which are formed at the estuaries, where freshwater and saltwater are merged, are distinct and dynamic. In addition to being vital centers for aquaculture, fisheries, agriculture, and other human endeavors, these areas are hotspots for biodiversity (Elliott et al. 2022). They are quite susceptible to changes in the environment, nevertheless, because of their temporary appeal. Water quality in these regions is greatly impacted by excess nutrients from intensive agriculture, urban runoff, industrial discharges, and human activity (Delkash et al. 2018). Moreover, estuaries are especially vulnerable to pollution and nutrient overloading since they are essential to delta ecosystems. For instance, pollution, salinity intrusion, and habitat loss pose serious problems for the Godavari Delta in India and the Mekong Delta in Vietnam (Nagaraju et al. 2023). Human-caused activities that upset the ecological balance of delta regions, like aquaculture, urbanization, and industrialization, worsen these problems (Bogardi et al. 2020).

A common economic activity in many delta locations is shrimp aquaculture, which serves as an example of the major concern between environmental sustainability and economic development (Neiland et al. 2001). Shrimp farming significantly influences the environment, even if it provides food security and livelihoods. High amounts of fertilizers, suspended particles, and organic matter in farmed effluents cause habitat damage, increased water turbidity, and oxygen depletion (Ferreira et al. 2011). For several months, the pond bottoms experience a total ammonia bloom from aquaculture effluents, which results in extremely toxic materials. For instance, the coast of northeastern Brazil, where shrimp cultivation has destroyed more than 30% of mangroves, interfered with natural drainage and increased saltwater intrusion (Barletta et al. 2023).

Delta ecosystems are even more vulnerable as a result of climate change. Due to climate change, rising sea levels have exacerbated the salinization of freshwater sources, making them unfit for industrial, agricultural, and residential use (Ondrasek and Rengel 2021). For example, between 1995 and 2020, Vietnam's sea level rose by about 14–15 cm, causing extensive saltwater intrusion in regions like Bac Lieu, Soc Trang, and the Ca Mau Peninsula (Nhung et al. 2019). In addition to compromising the water supply, these changes risk delta areas' ecological and financial viability.

In recent years, there has been an urgent need for efficient water quality control in delta ecosystems. Understanding ecosystem dynamics, measuring source links, and creating reliable indices for decision-making must be the main goals of scientifically sound management practices. A multidisciplinary study combining ecological, hydrological, and socioeconomic viewpoints is necessary for this approach (Wesselink et al. 2017). Determining baseline conditions and comprehending the variables controlling their operation is essential in managing delta ecosystems. Ecosystem structural and functional analysis can illuminate important mechanisms like biological productivity, hydrological flow, and nutrient cycling. Developing conceptual models that mimic the effects of different management measures on water quality requires this knowledge (Nagaraju et al. 2024a).

Indicators of ecosystem health and quantitative water quality are essential for tracking and directing management choices. Water quality is frequently evaluated using parameters such as electrical conductivity, salinity, pH, dissolved oxygen levels, and concentrations of nitrogenous compounds (Vamsi Nagaraju and Sunil 2024). Simplifying intricate ecological data into useful insights is still difficult. Interdisciplinary approaches to water quality evaluation and management are necessary to address the complex issues delta ecosystems face (Jensen and Morita 2020).

This chapter explores delta ecosystems' environmental issues and water quality dynamics. Moreover, it highlights the sustainable management strategies that guarantee ecological integrity and resource availability for future generations by examining the effects of aquaculture, human activity, and climate change on water quality.

1.2 South Asia's Godavari and Mekong Deltas

The Godavari Delta in India and the Mekong Delta in Vietnam are two of the world's most economically and environmentally significant delta regions, supporting agriculture, fisheries, and aquaculture while showcasing diverse habitats. However, due to climate change and human activities, these regions are more affected by environmental issues. The water quality in the Godavari Delta has been severely affected by contaminants from untreated sewage, industrial discharges, and agricultural runoff, leading to eutrophication, oxygen depletion, and algal blooms. Furthermore, despite the financial advantages, the rapid expansion of shrimp aquaculture has resulted in saline intrusion, habitat destruction, and increasing pollution. The Mekong Delta also faces similar problems with salt intrusion caused by sea level rise and upstream dam building, which reduces freshwater flow (Tran et al. 2021). The region's aquaculture industry makes a significant economic contribution to the community. On the other hand, it generates nutrient-rich effluent that destroys mangrove habitats and degrades water quality, weakening the delta's resistance to environmental pressures. The famous delta areas of South Asia are depicted in Fig. 1.1.

The Godavari Delta and Mekong Delta promote intensive aquaculture farming, which plays a complex relationship between natural ecosystems and human activity. Aquaculture effluents, habitat loss, salinity intrusion, and climate change negatively

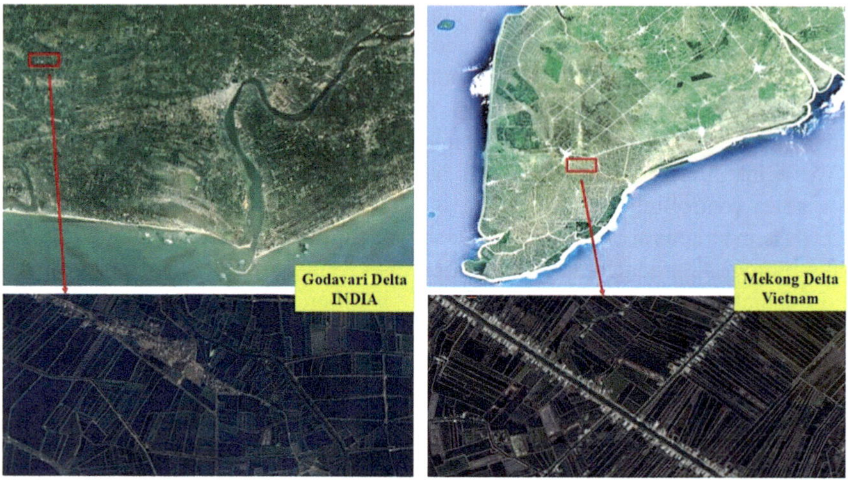

Fig. 1.1 Godavari Delta and Mekong Delta in South Asia

impact the water quality and biological balance of these regions (Nagaraju et al. 2022). To counteract this, an integrated and interdisciplinary approach to water quality management is required. Furthermore, sustainable management initiatives such as digital transformation and advanced tools are encouraged. Contaminants can be decreased by reducing chemical pesticides and fertilizers, promoting organic farming, and utilizing ecologically friendly aquaculture methods. To maintain biodiversity and strengthen the deltas' natural resistance to environmental changes, wetlands, mangroves, and other essential ecosystems also have positive potential.

Education and community involvement are essential for efficient delta management. Involving local communities in conservation initiatives guarantees that management techniques are socially and culturally acceptable. Policymakers must also consider the socioeconomic effects of conservation initiatives to reconcile ecological sustainability with economic development. Maintaining these essential ecosystems will guarantee their continuous contribution to regional economies, biodiversity, and the welfare of millions of people.

1.3 Water Quality Dynamics in Delta Ecosystems

Water quality is a critical determinant of ecological health and public well-being. However, many regions worldwide face significant threats to water quality due to a combination of human activities and the impacts of climate change. Despite increasing awareness, the mechanisms governing changes in water quality remain complex and not fully understood, making this a focal area of research in water sciences. Delta regions, characterized by intricate networks of rivers and streams, have experienced extensive degradation due to intense anthropogenic pressures. Globally, due to urban expansion and industrial development, approximately 60% of river systems have undergone structural changes (Kuenzer and Renaud 2012). Additionally, artificial barriers like dams have disrupted hydrological connectivity within these networks, compromising their self-purification capabilities and overall ecological health.

The interplay between water quality and environmental factors has attracted significant attention. However, few studies have deeply explored the intricate relationships between river systems and water quality, particularly in delta regions. River network structure and connectivity, key indicators of human influence, play a vital role in determining water quality through their impact on hydrological and biogeochemical processes (Covino 2017). For instance, in the Godavari Delta, elevated levels of total Kjeldahl nitrogen, which includes organic nitrogen and ammonia, have been recorded. Moreover, river density and morphology strongly correlate with water quality indices such as dissolved oxygen, chemical oxygen demand, and total phosphorus levels. River connectivity, encompassing structural and functional aspects, significantly influences the transport and distribution of pollutants. Structural connectivity refers to the continuity of river channels, while functional connectivity describes the interactions between these channels and catchment processes that enable water transfer. Research indicates that higher hydrological connectivity enhances the distribution of dissolved oxygen, as seen in connected wetland systems compared to isolated ones. Table 1.1 shows the water quality dynamics in the various delta ecosystems globally.

From global delta regions, human interventions that change the way pollutants are transported in delta regions are frequently disregarded. These structures, which were mostly built for environmental management and flood control, often interfere with water cycles, decrease connection, and impede aquatic species movement, which results in extreme water pollution occurrences. Rapid industrialization in China has severely deteriorated the quality of the water in areas like the Pearl River Delta and the Yangtze River Delta (Yi et al. 2003). Because of ongoing pollution, the Yangtze River Delta, a level plain with extensive water networks, experiences acute water shortages every year. Developing ecological strategies and managing water resources sustainably require an understanding of how river structure and connections affect water quality.

Due to land-use changes, industrial expansion, and competing stakeholder interests, delta regions, which are frequently low-lying and biologically sensitive, are becoming more and more susceptible to environmental deterioration. These elements, which are made worse by global sea level rise, lead to habitat degradation, coastal erosion, and increased flood hazards. Sea level rise, for instance, is a further threat to the sustainability of the mangrove ecosystems along India's eastern coast, which are particularly vulnerable to these stresses. A thorough grasp of the relationships between hydrological connectivity, human activity, and water quality is necessary to address the complex issues facing delta ecosystems and to create sustainable management and conservation plans.

Table 1.1 Delta ecosystems' water quality challenges

Delta and country	Challenges	Major water quality parameters	References
Godavari Delta, India	Impacted by industrial discharges, agricultural runoff, and aquaculture effluents; increasing salinity, nutrient overloading, and sedimentation are significant concerns	Total nitrogen, phosphorus, dissolved oxygen, biochemical oxygen demand, and salinity levels	Nagaraju et al. (2024b)
Mekong Delta, Vietnam	Salinity intrusion, pesticide contamination, and sediment transport reduction due to upstream dam construction	Salinity, nitrate levels, dissolved oxygen, and suspended sediment concentrations	Clauss et al. (2018)
Yangtze Delta, China	Rapid urbanization and industrialization	Chemical oxygen demand, ammonium nitrogen, heavy metals, and total phosphorus	Yi et al. (2003)
Nile Delta, Egypt	Reduced freshwater flow, salinity intrusion, pesticide contamination, and untreated sewage	Salinity, pesticide residues, and nitrogenous compounds	Nour et al. (2022)
Don Delta, Russia	Pollution from agricultural runoff, industrial discharges, and reduced water flow due to damming	Phosphates, pH, and heavy metal concentrations	Venevsky et al. (2024)
Yellow River Delta, China	Reduced river flow, high sediment load, and industrial and agricultural pollutants	Turbidity, nitrogen, phosphorus, and biochemical oxygen demand	Zhang et al. (2022)
Niger Delta, Nigeria	Oil spills, illegal refineries, and poor waste management	Hydrocarbons, heavy metals, and dissolved oxygen levels	Okoye et al. (2022)
Cauvery Delta, India	Agricultural runoff, industrial pollution, and reduced water flow due to upstream water diversion	Total dissolved solids, nitrogen, phosphorus, and dissolved oxygen	Gottumukkala et al. (2024)
Pearl Delta, China	Eutrophication, heavy metal contamination, and sediment imbalance	Heavy metals (e.g., lead, mercury), biochemical oxygen demand, and total nitrogen	Zhang et al. (2021)

1.4 Research Focus on Delta Ecosystems

Agriculture depends on delta areas with rich soils and an abundance of water resources supported by intricate canal networks and delta habitats. In the past, these regions have been crucial to preserving agricultural productivity and food security. However, expanding urbanization, industry, and aquaculture operations have caused major environmental problems in recent years in important delta regions like the Mekong Delta in Vietnam and the Godavari Delta in India. Due to habitat destruction, water pollution, salinity intrusion, and biodiversity loss, these activities pose a major danger to ecological and economic viability.

The growing environmental problems in delta ecosystems have attracted a lot of scholarly interest. Scientists and politicians are increasingly focused on comprehending and mitigating the environmental impacts of human activity in these regions. This emphasis on research is reflected in the rise in publications in recent years. Over 800 research articles on delta ecosystems have been published in the last 10 years, per a survey of the Scopus database. The majority of these studies originate from China, India, and Vietnam, which are home to some of the world's most significant delta regions.

To find trends and research focus areas in delta ecosystems, a co-occurrence analysis of phrases was conducted using VOSviewer. Our analysis clarified the primary themes and connections throughout the study's corpus, as illustrated in Fig. 1.2. Among the primary topics were habitat restoration, sustainable agriculture, water quality, and the effects of climate change. These studies focused especially on the Mekong Delta and Godavari Delta, highlighting their global significance and the urgent need to address their environmental problems. As delta regions suffer from the effects of human activity and climate change, research on sustainable management and restoration is essential. Our understanding of these dynamic ecosystems is expanded by these studies, which also help shape laws and practices that strike a balance between economic growth and environmental preservation.

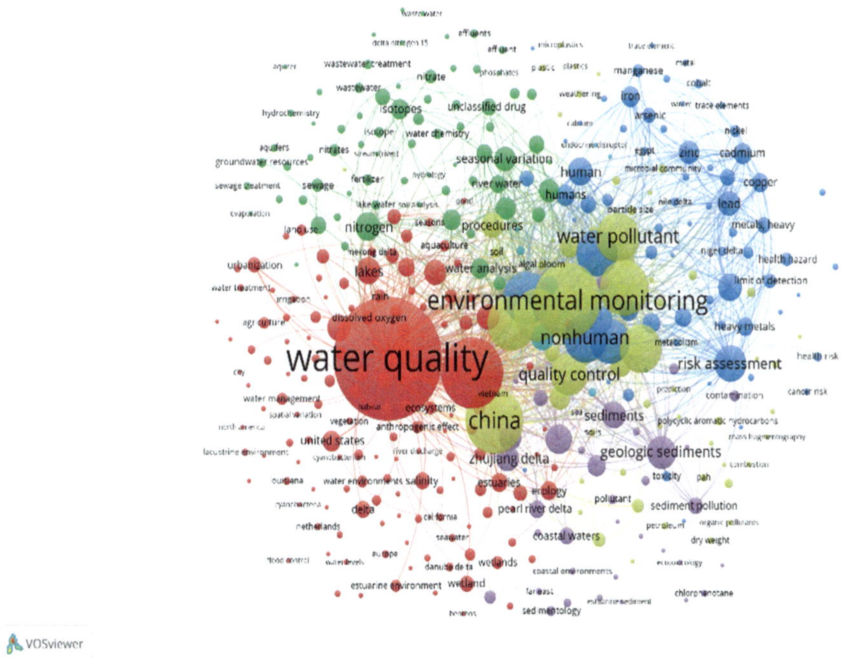

Fig. 1.2 Keyword co-occurrence of research articles visualized using VOSviewer

1.5 Conclusions

Delta ecosystems are highly complex and dynamic systems that get their water from various sources, such as rivers, canals, and sewers. These systems support aquaculture, agriculture, livelihoods, and biodiversity. Several physical, chemical, and bacteriological factors influence the quality of the water, which is an essential part of delta ecosystems. Key indicators such as water temperature, dissolved oxygen, nitrogen compounds, pH, alkalinity, salinity, conductivity, carbon dioxide levels, and turbidity are essential for evaluating the health of these ecosystems.

Chemical and biological oxygen demand, dissolved oxygen, and nutrient levels all significantly impact the quality of water in delta areas worldwide. High salt levels threaten the ecosystem's equilibrium and freshwater supply, and harmful pollutants further endanger aquatic life's and human populations' health. Human activities, including urbanization, industrialization, intensive agriculture, aquaculture, and climate change, which modify hydrological patterns and speed up saline intrusion, all contribute to these problems.

To address these issues, area-specific approaches that take into account the distinct environmental, social, and economic circumstances of each delta are needed. Restoring hydrological connections, encouraging sustainable resource use, and reducing pollution sources should be the top priorities of practical management strategies. Achieving long-term sustainability in these subtle ecosystems requires combining scientific study, legislative efforts, and community involvement.

Maintaining the health and functionality of delta ecosystems is essential for human well-being, global environmental stability, and the regions themselves. It is feasible to address the urgent issues these dynamic systems face and guarantee their resilience for future generations by utilizing interdisciplinary research and creative approaches to economic growth and environmental protection.

References

Akhtar N, Syakir Ishak MI, Bhawani SA, Umar K (2021) Various natural and anthropogenic factors responsible for water quality degradation: a review. Water 13(19):2660

Barletta M, Melo RC, Whitfield AK (2023) Past and present conservation of south American estuaries. Estuarine Coastal Shelf Sci 295:108542

Bogardi JJ, Leentvaar J, Sebesvári Z (2020) Biologia Futura: integrating freshwater ecosystem health in water resources management. Biol Futura 71(4):337–358

Clauss K, Ottinger M, Leinenkugel P, Kuenzer C (2018) Estimating rice production in the Mekong Delta, Vietnam, utilizing time series of Sentinel-1 SAR data. Int J Appl Earth Obs Geoinf 73:574–585

Covino T (2017) Hydrologic connectivity as a framework for understanding biogeochemical flux through watersheds and along fluvial networks. Geomorphology 277:133–144

Delkash M, Al-Faraj FA, Scholz M (2018) Impacts of anthropogenic land use changes on nutrient concentrations in surface waterbodies: a review. CLEAN – Soil Air Water 46(5):1800051

Elliott M, Houde ED, Lamberth SJ, Lonsdale JA, Tweedley JR (2022) Management of fishes and fisheries in estuaries. Fish Fish Estuaries Glob Perspect 2:706–797

Ferreira NC, Bonetti C, Seiffert WQ (2011) Hydrological and water quality indices as management tools in marine shrimp culture. Aquaculture 318(3–4):425–433
Giri S (2021) Water quality prospective in twenty first century: status of water quality in major river basins, contemporary strategies and impediments: a review. Environ Pollut 271:116332
Gottumukkala SB, Thotakura VN, Gvr SR, Chinta DP, Park R (2024) Balancing aquaculture and estuarine ecosystems: machine learning–based water quality indices for effective management. Environ Sci Pollut Res:1–17
Jensen CB, Morita A (2020) Deltas in crisis: from systems to sophisticated conjunctions. Sustain For 12(4):1322
Kuenzer C, Renaud FG (2012) Climate and environmental change in river deltas globally: expected impacts, resilience, and adaptation. In: The Mekong delta system: interdisciplinary analyses of a river delta. Springer, Dordrecht, pp 7–46
Lim YA, Nissapatorn V (2017) Transmission of waterborne parasites in the Association of Southeast Asian Nations (ASEAN): overview and direction forward. Food Waterborne Parasitol 8:75–83
Mukherjee S, Patel AK, Kumar M (2020) Water scarcity and land degradation nexus in the Anthropocene: reformations for advanced water management as per the sustainable development goals. In: Emerging issues in the water environment during Anthropocene: a South East Asian perspective, pp 317–336
Nagaraju TV, Malegole SB, Chaudhary B, Ravindran G (2022) Assessment of environmental impact of aquaculture ponds in the western delta region of Andhra Pradesh. Sustain For 14(20):13035
Nagaraju TV, Sunil BM, Chaudhary B, Prasad CD, Gobinath R (2023) Prediction of ammonia contaminants in the aquaculture ponds using soft computing coupled with wavelet analysis. Environ Pollut 331:121924
Nagaraju TV, Bala GS, Bonthu S, Mantena S (2024a) Modelling biochemical oxygen demand in a large inland aquaculture zone of India: implications and insights. Sci Total Environ 906:167386
Nagaraju TV, Malegole SB, Chaudhary B, Ravindran G, Chitturi P, Chinta DP (2024b) Novel assessment tools for inland aquaculture in the western Godavari delta region of Andhra Pradesh. Environ Sci Pollut Res 31(25):36275–36290
Neiland AE, Soley N, Varley JB, Whitmarsh DJ (2001) Shrimp aquaculture: economic perspectives for policy development. Mar Policy 25(4):265–279
Nhung TT, Le Vo P, Van Nghi V, Bang HQ (2019) Salt intrusion adaptation measures for sustainable agricultural development under climate change effects: a case of Ca Mau Peninsula, Vietnam. Clim Risk Manage 23:88–100
Nour HE, Alshehri F, Sahour H, El-Sorogy AS (2022) Evaluation of sediment and water quality of Ismailia Canal for heavy metal contamination, Eastern Nile Delta, Egypt. RegStud Mar Sci 56:102714
Ogidi OI, Akpan UM (2022) Aquatic biodiversity loss: impacts of pollution and anthropogenic activities and strategies for conservation. In: Biodiversity in Africa: potentials, threats and conservation. Springer, Singapore, pp 421–448
Okoye EA, Ezejiofor AN, Nwaogazie IL, Frazzoli C, Orisakwe OE (2022) Heavy metals and arsenic in soil and vegetation of Niger Delta, Nigeria: ecological risk assessment. Case Stud Chem EnvironEng 6:100222
Ondrasek G, Rengel Z (2021) Environmental salinization processes: detection, implications & solutions. Sci Total Environ 754:142432
Takyi SA, Amponsah O, Darko G, Peprah C, Apatewen Azerigyik R, Mawuko GK, Awolorinke Chiga A (2022) Urbanization against ecologically sensitive areas: effects of land use activities on surface water bodies in the Kumasi Metropolis. Int J Urban Sustainable Dev 14(1):460–479
Tran TA, Dang TD, Nguyen TH (2021) Moving towards sustainable coastal adaptation: analysis of hydrological drivers of saltwater intrusion in the Vietnamese Mekong Delta. Sci Total Environ 770:145125

Vamsi Nagaraju T, Sunil BM (2024) Briefing: intensive inland aquaculture ponds: challenges and research opportunities. Environ Geotech 40(XXXX):1–5

Venevsky S, Berdnikov S, Day JW, Sorokina V, Gong P, Kleshchenkov A et al (2024) Don River Delta hydrological and geomorphological transformation under anthropogenic and natural forcings. Dynamics of delta shoreline, risk of coastal flooding and related management options. Ocean Coast Manag 258:107364

Wesselink A, Kooy M, Warner J (2017) Socio-hydrology and hydrosocial analysis: toward dialogues across disciplines. Wiley Interdiscip Rev Water 4(2):e1196

Yi S, Saito Y, Zhao Q, Wang P (2003) Vegetation and climate changes in the Changjiang (Yangtze River) Delta, China, during the past 13,000 years inferred from pollen records. Quat Sci Rev 22(14):1501–1519

Zhang Z, Peng J, Xu Z, Wang X, Meersmans J (2021) Ecosystem services supply and demand response to urbanization: a case study of the Pearl River Delta, China. Ecosyst Serv 49:101274

Zhang CY, Zhao L, Zhang H, Chen MN, Fang RY, Yao Y et al (2022) Spatial-temporal characteristics of carbon emissions from land use change in Yellow River Delta region, China. Ecol Indic 136:108623

Chapter 2
Climate Change and Coastal Hazards in Delta Ecosystems

T. Vamsi Nagaraju and G. Sri Bala

Abstract Sea level rise and the escalation of catastrophic events brought on by climate change are two significant issues delta ecosystems must immediately face. Communities along the coast could suffer serious harm due to coastal hazards. However, several other climate-related risks also present immediate dangers. By addressing different deltas worldwide, this chapter investigates the effects of climate change on delta ecosystems. According to recent studies, rising greenhouse gas concentrations are bringing about conditions in delta systems that have not been observed in millions of years. This could lead to significant and permanent ecological changes. Due to anthropogenic climate change, delta production has decreased, food web dynamics have altered, habitat-forming species are less prevalent, species ranges have shifted, and disease prevalence has increased. We give a general overview of how human activity and global warming interact along the shore. This chapter aims to illustrate the importance of long-term monitoring, the development of ecological indicators, and the applications of knowledge on the response of aquatic ecosystem functioning and biodiversity to global warming. The findings can be applied to other regions. By including the perspectives of different deltas from around the world, this chapter gives fresh ideas and a distinctive viewpoint on the intricate nature of climate change's effects on delta ecosystems.

Keywords Delta region · Inland aquaculture · Ecosystem · Climate change · Salinity intrusion

T. V. Nagaraju · G. S. Bala (✉)
Department of Civil Engineering, SRKR Engineering College, Bhimavaram, Andhra Pradesh, India

Centre for Sustainable Inland Aquaculture, SRKR Engineering College, Bhimavaram, India

T. V. Nagaraju, G. S. Bala (eds.), *Delta Ecosystems*, Environmental Contamination Remediation and Management, https://doi.org/10.1007/978-3-031-97920-0_2

2.1 Introduction

Delta ecosystems may be seriously threatened by concerns related to climate change. Climate change directly threatens delta ecosystems through anticipated significant changes, rising sea levels, and rising temperatures (Scown et al. 2023). In addition to reducing water supplies and destroying agricultural land, these effects could negatively influence the nation's ecology, economy, and public health due to rising temperatures (Santosh et al. 2024). Many nations have been studying climate change worldwide for more than 20 years (Gupta et al. 2024). River deltas along the coast are areas of concern for sustainable development worldwide. Less than 1% of the Earth's land cover is inhabited by more than 5% of the world's inhabitants in deltas (Wei et al. 2024). Some South Asian deltas have metropolitan areas with population densities of well over 10,000 people per square kilometer, while other deltas are rapidly becoming more urbanized (Youssef et al. 2024). This makes it even more important to direct wise urban growth in these areas. Due to their extremely fertile vegetation fields and the availability of resources with dense waterbodies, deltas are also crucial locations for global trade and food security (Vamsi Nagaraju and Sunil 2024). However, the sustainability of these environmental goods flows is in jeopardy. However, because of their low height, deltas are particularly vulnerable to flooding, salinization, and relative sea level rise (Khan 2024). These hazards and additional urbanization are causing delta ecosystems to deteriorate or disappear. The necessity of accomplishing the United Nations Sustainable Development Goals (SDGs) in these regions and worldwide is strongly supported by the significance of deltas for progress and the significant environmental dangers (Scott 2024). There are several levels of environmental change that are important for long-term delta sustainability. The rise of sea level puts strain on these ecosystems worldwide. Drainage catchments, dams for water use, and hydropower expansion cause the delta to become disconnected from its upland region, forming a sediment load. Local effects of human activity within deltas include development, land drainage, and subsidence brought on by the extraction of hydrocarbons and groundwater (Gottumukkala et al. 2024). Many deltas are now "locked-in" to high-risk states and require expensive, inflexible, and energy-intensive adaptation measures due to the exploitation of deltas for social and economic gain (Do et al. 2024). Many deltas' socioeconomic structures are no longer tied to the ecosystems that are the foundation of their existence (Meng et al. 2024). These stresses and conditions inevitably impact the sustainability of deltas. Future changes in deltas worldwide under many situations and across balances remain very undefined and unmapped, even though the significance of deltas for sustainable growth is well unstated, and researchers have proposed measurements of delta hazard and sustainability (Li et al. 2024). Prior evaluations of future risk to deltas around the world have been conducted from a single angle, such as population exposure to flooding, which ignores other aspects of the delta ecosystem that influence risk, or by developing crude future situations in which individual energy prices alteration, failing to account for alternative paths of population development, land-use variation, sea-level escalation, or socioeconomic expansion over the next century. Integrated assessment models have been employed locally in specific deltas to analyze trade-offs under various future scenarios, but they have not yet been widely applied to many deltas worldwide (Van et al. 2024). Delta risk and sustainable development are both impacted by the scale problem. Here, we begin bringing the global to the local in deltas. Regardless of which of the potential outcomes materializes, local pressures on deltas will be imposed by global environmental change and socioeconomic development. Information on these changes' pressures on deltas is needed to define boundary settings for delta-scale situations and models (Feng et al. 2024). Advances in various ecological and social variables have been simulated and assessed worldwide for various situations. Global adjustments should establish boundary conditions reliable with global situations across all deltas rather than specific to each delta. However, if difficulties in deltas are not adequately addressed, they may worsen and impact global development and the SDGs' accomplishment (Lopes 2024). As a result, it is crucial to evaluate potential developments and identify dangers in deltas worldwide under various situations.

2.2 Climate Change Impact on Delta Ecosystems

Climate change mainly impacts delta ecosystems through wind patterns, precipitation, and air temperature modifications (Eid et al. 2024). Freshwater systems' reactions to physical changes, like the stratification and mixing regimes of lake water columns or the watershed, can lead to chemical changes in habitats, including oxygen content, nutrient cycling, and water color. Many lakes in tropical and subtropical regions see warmer water evaporating more quickly when global mean surface temperatures rise. Important riparian regions needed for seed dispersal and reproduction are in danger due to declining lake levels and erratic changes in water flow. Some canals or lakes may have saline or endorheic streams that support delta ecosystems (Nagaraju et al. 2022). Because primary productivity has dropped by 70% since 2000, there are fewer nutrients in the lake's upper layers, and the water is more transparent, making it easier for light to enter. In addition to having more notable seasonal temperature fluctuations than tropical lakes, temperate lakes also exhibit extreme thermal gradients (Xu et al. 2024). An essential component of the river's ecology is its flow regime. Seasonal variations in flow are essential to freshwater ecosystems, but variations in normal precipitation, evaporation, moisture levels in the soil, and groundwater recharge can change a mean flow (Sharma and Naik 2024). Figure 2.1 shows climate change impact on delta ecosystems.

When the global average temperature rises, most wetlands will dry up because the precipitation rate cannot keep up with the evaporation rate (McCullough 2024). If rainfall decreases and groundwater is taken out for human use, there will not be any traces of temporary wetlands. As the rainfall increased, wetland habitat was

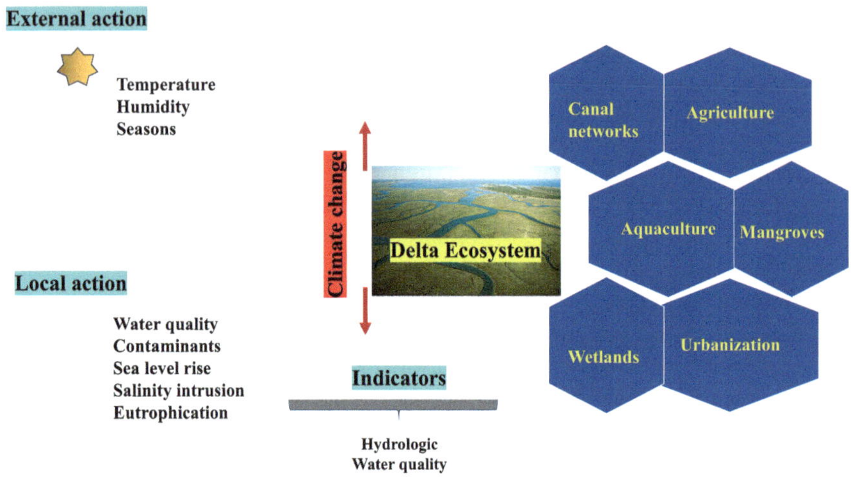

Fig. 2.1 Climate change impact on delta ecosystems

inundated, deepened, and expanded. However, flooding can destroy vegetation and other ecological characteristics in wetland ecosystems, and heavy rains increase the likelihood that silt and pollution will reach streams (Nagaraju et al. 2024a). Additionally, changing temperatures might cause unforeseen changes in the hydrological regime of glacial and subarctic regions. Drying up transitory wetland habitats can lead to the extinction of endangered species. For example, several native shrimp species in South Asian nations have lost their homes in shallow, ornate pools due to reduced rainfall and high evaporation (Dory et al. 2024). This is due to a decrease in the quantity and size of ponds and an increase in the distances between them. Because wetlands are an endangered habitat type, migratory birds rely on them as part of their migration route, and regional habitat connectivity is declining. Because they lessen wetlands' susceptibility to negative impacts, increased wetland acreage and accessibility may also benefit regions with heavy rainfall. The effects of climate change on aquatic life in delta habitats are presented in Table 2.1.

Human activity has significantly increased the nutrients that reach marine areas from delta ecosystems. However, due to a projected net decline in global ocean productivity, climate change negatively impacts the open ocean's near-surface waters. Wherever it is needed, nutrient recycling from the deep ocean into the surface ocean is essential for phytoplankton production, an essential component of marine productivity. Because changes in the productivity and composition of communities close to the ocean's surface affect the quantity and quality of organic matter carried to the ocean's interior and seafloor communities, decreases in the net production of primary substances substantially affect higher trophic levels.

Table 2.1 Effect of climate change on aquatic habitat

Parameters	Impact	Reference
Delta temperature and humidity rise	Temperature variations can change the water's nutritional composition and distribution. The quantity of biomass produced by phytoplankton is mostly determined by temperature and the availability of nutrients. Furthermore, zooplankton development tends to increase with temperature.	Boyd et al. (2025)
Acidification	When surface water's pH drops due to atmospheric carbon dioxide and CO_2 accumulation, delta acidification occurs. As CO_2 levels rise, water quality and productivity will decline.	Igwaran et al. (2024)
Eutrophication	Changes in the climate have also been connected to harmful algal blooms. Algal blooms seriously threaten the long-term sustainability of aquaculture and fishing operations in terms of environmental sustainability.	Nagaraju et al. (2024b)
Sea level rise	The salinity of the water and the availability of food in the environment impact the number and type of zooplankton and phytoplankton. Salinity changes may affect the ecosystem's taxonomic composition, nutrient cycles, decomposition, and food web efficiency.	Ghosh and Panigrahi (2024)
Anthropogenic	Because of human challenges, such as chemical spills, aquatic ecosystems have been harmed and have seen a major decline in function. Because of substantial anthropogenic pollution, such as runoff from the industrial and agricultural sectors and household sewage, flowing areas, such as rivers, are the most vulnerable environments.	Nagaraju et al. (2023)

2.3 Delta Ecosystems: Habitat Loss and Adaptation Strategies

The rise of sea level puts strain on these ecosystems worldwide. At the drainage basin level, dams for water use and hydropower development cause the delta to become disconnected from its upstream region, starving it of river sediment. Local effects of human activity within deltas include development, land drainage, and subsidence brought on by the extraction of hydrocarbons and groundwater. Many deltas are now locked into high-risk states and require expensive, inflexible, and energy-intensive adaptation measures due to the exploitation of deltas for social and economic gain (Yan et al. 2024). Many deltas' socioeconomic structures are no longer tied to the ecosystems that are the foundation of their existence (Abdelhamid et al. 2024). These stresses and conditions inevitably impact the sustainability of deltas.

The coastal habitats are being devastated and degraded in many places, even though maintaining healthy ecosystems that function effectively is crucial to optimizing their efficiency in lowering human susceptibility and promoting climate change variation. Around 75% of the global coral reefs are classified as vulnerable, 85% of the reefs containing oysters have disappeared worldwide, and local, regional, and international assessments indicate equally concerning levels of influence throughout all coastal ecosystems (Mondal et al. 2024). Mangrove habitats are disappearing at a rate of 0.7% annually, 3–4 times faster than the loss of forests worldwide (Zhong et al. 2024). Both direct and indirect effects pose a threat to coastal ecosystems. These include habitat loss or destruction, particularly due to land claims, aquaculture conversion, and mangrove harvesting, as well as damaging and overfishing, all of which have the potential to significantly alter the structure and function of ecosystems. Numerous land-based industrial and agricultural developments impacting the sediment, nutrient, and pollution levels in coastal waterways are considered indirect or ex-situ impacts. Changes in freshwater runoff and sediment delivery are often caused by converting natural lands to agriculture and subpar agricultural practices. Dams and intensive irrigation can lower freshwater flows. Coastal ecosystems are also impacted by climate change. Every coral reef region has already seen widespread coral bleaching, and coral and shellfish reefs may be seriously threatened by ocean acidification. The extraordinary losses in surface complexity, or rugosity, caused by the coral cover reduction in African reefs since early 2000 are believed to have diminished the reefs' ability to protect the coast in many places. As sea levels rise, continued urbanization is particularly concerning since it hinders coastal wetlands' natural migration inland and causes habitat loss through the coastal squeeze (Singha et al. 2024).

Social and economic vulnerability assessments have frequently only looked at the direct effects of climate change. This entails determining whether populations in coastal areas may be affected by coastal storms and sea level rise. Furthermore, the degree of social, economic, and cultural reliance on delta ecosystems and natural resources for work, income, enjoyment, and food should be taken into account in these evaluations. Future changes, such as anticipated human effects from climate change and other factors such as social, demographic, and economic shifts brought about by the travel and fishing industries, residential or industrial growth patterns, and similar factors, should be considered in vulnerability assessments. Such a shift in the future is likely to significantly impact the regional patterns and temporal trends of vulnerability to sea level rise. In keeping with this, vulnerability assessments that consider coastal engineering solutions must shift from a crude evaluation of construction costs to a comprehensive analysis of the ripple effects on entire coastlines and the losses of ecosystem amenities for societies as ecosystems are wedged.

Due to their rarity or future nature, delta threats and climate change can be challenging to prepare. Furthermore, many important coastal choices are decided locally, with little access to planning tools. More and more tools and decision-support systems are being created to assist communities in visualizing the socioeconomic and ecological effects of climate change and coastal hazards and thinking through potential remedies. These tools frequently include location-specific storm danger and sea level rise forecast climate change models (Bernier et al. 2024). They can assist communities in evaluating social and economic vulnerability and risk. For instance, they can assist in determining which individuals are least likely to escape danger and whose vital infrastructure is most at risk. More significantly, they can assist communities in assessing options to lower the risk, vulnerability, and expenses related to climate change and natural calamities. Plans for restoration, hazard management, land use zoning, coastal development planning, and other resource management strategies can all benefit from these solutions.

2.4 Conclusions

Despite abundant evidence supporting climate change, environmental changes and variations in greenhouse gas emissions into the atmosphere will always impact projections. Human actions like habitat destruction and land cover changes can also accelerate the negative consequences of climate change. Species will either locate adequate habitats, adapt to the new environment, or become extinct due to climate change. Climate change will harm most species, although if habitat is increased appropriately, some species might gain. Finding victors and losers is a major issue, according to studies on climate change. Delta habitats are impacted by several stressors caused by human-induced climate change. Evaluating how ecosystems react to different stresses in their natural environments is crucial since these factors may have additive, detrimental, or beneficial effects. Delta habitats are impacted by several stressors brought on by human-caused climate change. Because of the potential for additive, synergistic, or antagonistic effects of various variables, ecosystem responses must be examined in their natural settings while considering all pertinent pressures. However, a person's susceptibility to environmental conditions predicted under climate change scenarios may be altered by the environmental conditions observed by previous generations and throughout early development. Future research has to consider this. It is difficult to predict how much the aquaculture sector will be impacted and how we could react to it because of the lack of emphasis on scientific research.

A thorough analysis of the most significant wetland services, such as water filtration and climate change mitigation, is necessary to propose sustainable and effective management. However, it will frequently be difficult to undertake trials to find such an approach, primarily due to time and money constraints. Using process-based models is one method to get beyond this obstacle. Assume that the many datasets produced by the experiment are adequate to depict the systems' pattern of change. In that scenario, a highly developed algorithm ought to be able to accurately identify and forecast the system's reaction while conserving funds, time, and resources. Machine learning techniques are perfect for simulating a delta environment because they can adapt to the complexity of the relationships and interactions between variables. Furthermore, multi-variable monitoring thoroughly explains the wetlands' positive and negative climate change feedback. This aids in identifying suitable management measures that environmental managers might employ to preserve delta ecosystems and slow climate change.

References

Abdelhamid S, Reinicke GB, Klaus R, Höhn J, Saad OS, Grenzdörffer G (2024) Red Sea coral reef monitoring site in Sudan after 39 years reveals stagnant reef growth, continuity and change. Diversity 16(7):379

Bernier NB, Hemer M, Mori N, Appendini CM, Breivik O, de Camargo R et al (2024) Storm surges and extreme sea levels: review, establishment of model intercomparison and coordination of surge climate projection efforts (SurgeMIP). Weather Clim Extremes 45:100689

Boyd CE, Davis RP, McNevin AA, Kumar V (2025) Water quality and its impacts on feeding practices. In: Feed and feeding for fish and shellfish. Academic, pp 383–401

Do DT, Nguyen HQ, Do XS (2024) Sustainable development in Mekong Delta, Vietnam: designing resilient cities based on landscape ecological principles. J Asian Archit Build Eng 23:1–18

Dory F, Nava V, Spreafico M, Orlandi V, Soler V, Leoni B (2024) Interaction between temperature and nutrients: how does the phytoplankton community cope with climate change? Sci Total Environ 906:167566

Eid MH, Eissa M, Mohamed EA, Ramadan HS, Czuppon G, Kovács A, Szűcs P (2024) Application of stable isotopes, mixing models, and K-means cluster analysis to detect recharge and salinity origins in siwa oasis, Egypt. Groundw Sustain Dev 25:101124

Feng S, Shafiei MWM, Ng TF, Ren J, Jiang Y (2024) The intersection of economic growth and environmental sustainability in China: pathways to achieving SDG. Energ Strat Rev 55:101530

Ghosh P, Panigrahi AK (2024) Assessment of water quality and source apportionment of pollution in a tropical river in eastern India: a study utilizing multivariate statistical tools and the APCS-MLR receptor model. Environ Monit Assess 196(9):861

Gottumukkala SB, Thotakura VN, Gvr SR, Chinta DP, Park R (2024) Balancing aquaculture and estuarine ecosystems: machine learning–based water quality indices for effective management. Environ Sci Pollut Res 31:1–17

Gupta A, Barik SK, Chauhan PS (2024) Towards sustainable delta ecosystems: pollution mitigation for achieving SDGs in Indian Delta Region. In: Solid waste management in Delta Region for SDGs fulfillment: Delta sustainability by waste management. Springer, Cham, pp 27–46

Igwaran A, Kayode AJ, Moloantoa KM, Khetsha ZP, Unuofin JO (2024) Cyanobacteria harmful algae blooms: causes, impacts, and risk management. Water Air Soil Pollut 235(1):71

Khan MA (2024) Analyzing three zeros (zero poverty, unemployment, and carbon emissions) in Asia and the Pacific region: assessment of sustainable development goals through the STIRPAT model. Sustain Dev 32:4782

Li T, McDermid SS, Valdivia RO, Sundaram P (2024) Climate change mitigation and adaptation for rice-based farming systems in the Red River Delta, Vietnam. CABI Agric Biosci 5(1):109

Lopes MC (2024) Climate change and its impact on children and adolescents sleep. J Pediatr 101:S40

McCullough CD (2024) Below water table mining, Pit lake formation, and management considerations for the Pilbara Mining Region of Western Australia. Mining 4(4):863–888

Meng C, Chen Y, Yang J, Su X, Guo W, Zhang K (2024) Association between multilevel landscape characteristics and rural sustainability: a case study of the water-net region in the Yangtze River Delta, China. Eco Inform 102677:102677

Mondal I, Naskar PK, Alsulamy S, Jose F, Hossain SA, Mohammad L et al (2024) Habitat quality and degradation change analysis for the Sundarbans mangrove forest using invest habitat quality model and machine learning. Environ Dev Sustain 26:1–26

Nagaraju TV, Malegole SB, Chaudhary B, Ravindran G (2022) Assessment of environmental impact of aquaculture ponds in the western delta region of Andhra Pradesh. Sustain For 14(20):13035

Nagaraju TV, Sunil BM, Chaudhary B, Prasad CD, Gobinath R (2023) Prediction of ammonia contaminants in the aquaculture ponds using soft computing coupled with wavelet analysis. Environ Pollut 331:121924

Nagaraju TV, Malegole SB, Chaudhary B, Ravindran G, Chitturi P, Chinta DP (2024a) Novel assessment tools for inland aquaculture in the western Godavari delta region of Andhra Pradesh. Environ Sci Pollut Res 31(25):36275–36290

Nagaraju TV, Bala GS, Bonthu S, Mantena S (2024b) Modelling biochemical oxygen demand in a large inland aquaculture zone of India: implications and insights. Sci Total Environ 906:167386

Santosh M, Groves DI, Yang CX (2024) Habitable planet to sustainable civilization: global climate change with related clean energy transition reliant on declining critical metal resources. Gondwana Res 130:220–233

Scott CA (2024) Land subsidence in the mid-Atlantic United States: creeping disaster threatens water, energy, and climate security. Water Secur 23:100183

Scown MW, Dunn FE, Dekker SC, van Vuuren DP, Karabil S, Sutanudjaja EH et al (2023) Global change scenarios in coastal river deltas and their sustainable development implications. Glob Environ Chang 82:102736

Sharma LK, Naik R (2024) Saline wetland ecosystems. In: Conservation of saline wetland ecosystems: an initiative towards UN decade of ecological restoration. Springer, Singapore, pp 33–75

Singha P, Dasgupta R, Hashimoto S, Mitra BK (2024) An integrated coastal exposure modelling approach to assist mangrove ecosystem based disaster risk reduction (Eco-DRR) in Tamil Nadu, India. Prog Disaster Sci 24:100388

Vamsi Nagaraju T, Sunil BM (2024) Briefing: intensive inland aquaculture ponds: challenges and research opportunities. Environ Geotech 40(XXXX):1–5

Van CM, Mei X, Mai TC (2024) Mekong River Delta. In: Delta sustainability: a report to the megadelta programme of the UN ocean decade. Springer, Singapore, pp 165–185

Wei Z, Tu J, Xiao L, Sun W (2024) Urbanization and carbon emissions in China: analysis of dynamic relationships from 1978 to 2020. J Geogr Sci 34(10):1925–1952

Xu X, Zhao J, Wu G, Li Y, Hou L (2024) Tracing water recharge and transport in the root-zone soil of different vegetation types in the Poyang Lake floodplain wetland (China) using stable isotopes. Sustain For 16(5):1755

Yan X, Han Z, Zou C, Cheng C (2024) Assessing the role of emerging green technology transfer in sustainable development and identification of key regions in Yangtze River Delta region. Technol Forecast Soc Chang 200:123099

Youssef YM, Gemail KS, Atia HM, Mahdy M (2024) Insight into land cover dynamics and water challenges under anthropogenic and climatic changes in the eastern Nile Delta: inference from remote sensing and GIS data. Sci Total Environ 913:169690

Zhong J, Liang C, Zhao Y, Wang Y, Yan X (2024) Identifying loss threshold and migration trajectory in the management of Suaeda salsa wetland under coastal squeeze. Mar Environ Res 194:106329

Chapter 3
Groundwater Dynamics of Godavari Delta in South India: A Geospatial Approach

G. Sri Bala and T. Vamsi Nagaraju

Abstract Inland aquaculture, once a driving force behind Andhra Pradesh's economic success, faces mounting challenges from climate change, including intensified farming practices, groundwater degradation, and salinity intrusion. Despite growing research on water quality in the Godavari Delta's aquaculture zones, limited attention has been paid to synthesizing findings or addressing environmental and societal sustainability. This chapter provides a comprehensive literature review focusing on high-density inland aquaculture's impact on groundwater quality and climate change. A case study on land use and groundwater dynamics in the Godavari–Krishna inter-delta is presented, employing geospatial techniques to assess climate change implications. The findings offer critical insights into the environmental consequences of intensive aquaculture and propose sustainable strategies applicable to other deltaic regions worldwide. This synthesis underscores the need for a balanced approach to harness the economic potential of aquaculture while safeguarding ecological and societal well-being.

Keywords Delta region · Inland aquaculture · Groundwater · Climate change

3.1 Introduction

Aquaculture is recognized as a crucial sustainable food production sector from an economic standpoint across the globe. Numerous low-income women and men depend on the fishing and aquaculture industries for income, which directly or indirectly enhances their food security (Dogeje et al. 2025). Intensive aquaculture operations sometimes require ample water consumption, leading to groundwater

G. S. Bala · T. V. Nagaraju (✉)
Department of Civil Engineering, SRKR Engineering College, Bhimavaram, Andhra Pradesh, India

Centre for Sustainable Inland Aquaculture, SRKR Engineering College, Bhimavaram, India

T. V. Nagaraju, G. S. Bala (eds.), *Delta Ecosystems*, Environmental Contamination Remediation and Management, https://doi.org/10.1007/978-3-031-97920-0_3

depletion and excess contamination from nutrients and pollutants. The elevated density and swift proliferation of these farms may also induce local climatic alterations, resulting in heat generation and ecosystem disruptions (John et al. 2024). Moreover, the accumulation of garbage and the application of antibiotics might further compromise water quality standards, endangering adjacent ecosystems and biodiversity as well (John et al. 2024). Disparately, problems such as high temperatures and summer dryness diminish water availability and pose a considerable danger to aquaculture systems. As a result, insufficient fresh water inflow leads to the concentration of metabolic byproducts such as CO_2 and NH_3 accumulation that might rapidly surpass acceptable limits (Ende et al. 2024). In particular, inland aquaculture fisheries are water-intensive enterprises that rely heavily on groundwater in addition to surface water. Excessive groundwater extraction induces sediment subsurface compaction, leading to land subsidence, or the reduction of the ground surface elevation (Ha et al. 2024). The significant reduction of groundwater hydraulic head results in seawater intrusion in the coastal aquifers, leading to the exhaustion of freshwater resources and an increase in socioeconomic expenses (Gottumukkala et al. 2024). Even, cation and anion pollution significantly impact groundwater variables and ecological balance. Elevated cation concentrations and heavy metals such as lead, mercury, and cadmium can arise from rapid industrialization, mining activities, and urban runoff. These contaminants can bio-accumulate in aquatic organisms, offering a threat to human and coastal ecosystems.

Besides, the significant diversity in species, cultural practices, feed types, and salt tolerance enhances the resilience of aquaculture to climate change; yet, it does not render aquaculture highly immune to disasters (Chengula et al. 2024). Severe weather phenomena and rising sea levels may result in flooding, water scarcity, and alterations in salinity, temperature, and dissolved oxygen levels. Increased disease prevalence may lead to greater losses and diminished animal well-being and impact the availability of external feed inputs for aquaculture (Tchonkouang et al. 2024). Aquaculture, on the other hand, exacerbates global warming due to heightened energy and feed requirements associated with intensification and greenhouse gas (GHG) emissions (Bashir et al. 2025). However, with rapid technological advancements, it may encompass energy conservation in fishing methods and aquaculture production, as well as enhanced post-harvest and distribution systems. Significant interactions may exist within the sector concerning environmental facilities, such as the preservation of coral reefs, coastal boundaries, and inland waterbodies, as well as potential carbon sequestration and nutrient balance strategies; however, further research and development is necessary in these cultured aspects.

On the other side, conventional groundwater monitoring techniques encounter difficulties in spatial representation and temporal resolution, frequently insufficient in coverage to depict the heterogeneity of groundwater systems and capture rapid fluctuations in groundwater dynamics (Dalu et al. 2024). The establishment of boreholes, manual measurements, and laboratory tests, although crucial, can be costly and resource-intensive. This budgetary constraint restricts the viability of developing comprehensive monitoring networks, especially in areas with scarce financial resources, and in due course, it hinders a thorough comprehension of the intricate dynamics of groundwater systems (Nagaraju et al. 2023).

Therefore, researchers employed various methodologies, including multivariate statistics, geochemical modeling, and the water quality index, to analyze and delineate the spatiotemporal variations of groundwater physicochemical properties. Additionally, Geographic Information Systems (GIS) is highly effective for the examination and analysis of spatial data pertaining to water resources. Improvements in sensor capabilities, spatial resolution, and data processing methods enhance the comprehension of groundwater dynamics and quality aspects (Zhan et al. 2024). Besides, the incorporation of artificial intelligence and machine learning significantly improves the precision of predictive models, providing essential tools for forecasting alterations in groundwater fluctuations. Besides, satellites outfitted with diverse sensors acquire photographs of the Earth's surface across many wavelengths, encompassing visible, infrared, and microwave spectra. They enable the identification of alterations in land cover and land use, offering significant insights into the processes influencing groundwater recharge.

Aircraft-mounted cameras acquire high-resolution photos of the Earth's surface, facilitating many uses in groundwater monitoring (Miyagi et al. 2024). These cameras provide meticulous mapping of terrestrial characteristics and hydrological conditions, yielding essential data for thorough evaluations. Furthermore, the technology facilitates the identification of subsurface geological structures, providing insights into the determinants affecting groundwater flow patterns (Ibrahim et al. 2024). The ability to see alterations in land use and urbanization via high-resolution imagery is essential for comprehending their effects on groundwater recharge. Besides, the advanced LiDAR systems employ laser beams to ascertain distances between the sensor and the Earth's surface, producing highly precise three-dimensional point cloud data. These systems are utilized in groundwater monitoring, specifically for the precise mapping of topography and land surface elevation (Jiang et al. 2024).

On the other side, freshwater supplies are susceptible to and may be significantly affected by climate change along with the quality criterion (Kim et al. 2024). The susceptibility of inland fisheries to climate change can be mitigated by adaptive measures that address unexpected outcomes and exploit potential opportunities (Roy et al. 2024). Governments must integrate these strategies into climate adaptation guidelines to ensure sustainability. Aquaculture experts, managers, and scientists should incorporate climate considerations into their work, including ethical reviews of angling practices during warm periods or closing overexploited fisheries to conserve stocks (Nagaraju et al. 2024a). Additional measures include diversifying livelihoods for vulnerable communities, managing fish stocks based on projected impacts, advancing technology, and planning for high-risk areas. Proactive, collaborative efforts are essential to protect inland fisheries and the communities that depend on them.

3.2 Groundwater Dynamics of Inland Aquaculture Zone of the Delta Ecosystem

3.2.1 High-Density Aquaculture Zones

Over 10% of global fish and shellfish species inhabit India's abundant and varied fisheries, encompassing deep seas, lakes, ponds, and rivers (Kumar et al. 2024). Currently, India accounts for 7.96% of global fish production (Naik 2024). India ranks among the leading seafood exporters globally, making the fishing industry vital to its economy and historically a substantial source of foreign exchange revenue.

There exist hundreds of freshwater and brackish water farmers development agencies dedicated to the advancement of freshwater and coastal aquaculture (Jaikumar et al. 2023). Significant species cultivated in India include Indian big carp and shrimp production. In addition, seaweed harvesting has gradually gained significance in the aquaculture landscape in recent years as a substitute for livelihood-supporting industries through small-scale practices. In this context, India's aquaculture resources comprise million hectares of inland aquaculture ponds, lakes, canals, and reservoirs available for aquaculture utilization (Nagaraju et al. 2024b). Ponds serve as the primary resources for both freshwater and brackish water aquaculture; however, only approximately 40% of landcover is utilized for farming. Ponds in eastern India are often earthen ponds measuring smaller than 1 ha, whereas the inland ponds on the east coast of India are larger, spanning areas of 15–25 ha each (Saizen et al. 2024). In northern India, open water bodies with inflows are prevalent, but in southern India, ponds known as tanks are primarily utilized for irrigation. In various regions of the country, ponds and tanks are either state-owned or community-based and are rented for durations of 2–10 years (Rana et al. 2024).

In recent decades, the Godavari–Krishna inter-delta region has experienced a notable land use transformation, with mangroves and paddy fields increasingly replaced by large-scale shrimp farming. This change, primarily motivated by economic benefits and the growing export demand, has resulted in soil salinization challenges. Farmers have resorted to intensifying shrimp cultivation, often at the expense of environmental sustainability. Figure 3.1 shows the intensity of inland aquaculture dominance in the inter-delta.

Table 3.1 summarizes the major state-level distribution of inland shrimp production across coastal regions of India, provided by MPEDA in 2023 (Pradhan et al. 2023). A state-wise analysis revealed a consistent relationship between shrimp production and detected aquaculture areas. Based on official data, states with extensive aquaculture coverage also recorded high shrimp production levels. However, exceptions like Tamil Nadu and Gujarat exhibited relatively low shrimp production compared to their detected aquaculture areas. Significant variations were observed among Indian states regarding aquaculture areas and shrimp production, influenced by differences in traditions, developmental strategies, and current statistics.

Andhra Pradesh stands out as the leading producer of brackish water shrimp, driven by economic reforms. It accounts for nearly 40% of India's total aquaculture

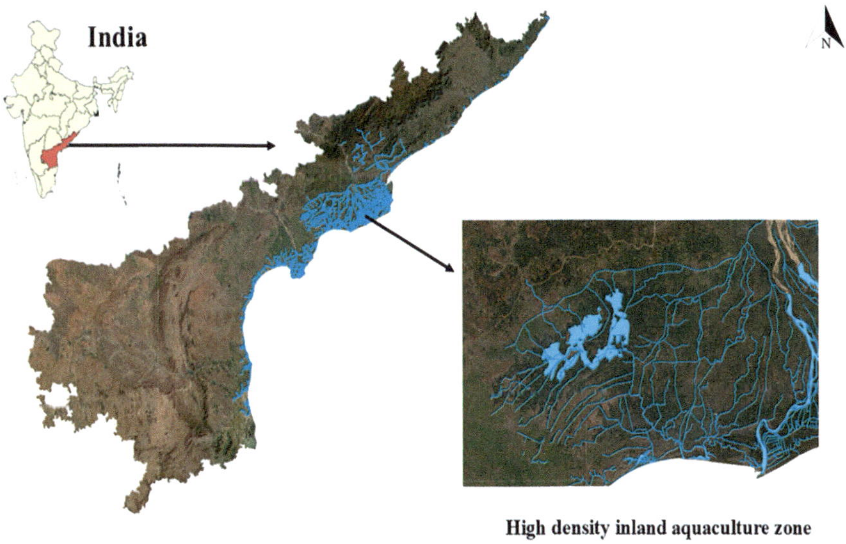

Fig. 3.1 High-density aquaculture zone in India

Table 3.1 Indian shrimp production state-wise in 2023 (Pradhan et al. 2023)

State	Shrimp production in metric tons
Gujarat	35,000
Kerala	5000
Tamil Nadu	23,000
Andhra Pradesh	600,000
Odisha	73,000
West Bengal	62,000

area and holds the largest share of ponds in the country (Pandey et al. 2024). At the district level, Krishna and West Godavari districts in Andhra Pradesh dominate aquaculture production. Figure 3.2 illustrates the detection of changes in aquaculture ponds within these districts. In this study, the land use land cover dynamics were performed from the Sentinel 2A data using ArcGIS software. From land use land cover, a 3% increase in water bodies and a 2.5% decrease in agricultural lands were found.

3.2.2 *Groundwater Usage Patterns in Inland Aquaculture*

Health issues are frequently linked to prolonged consumption of contaminated water, with pollutants arising from natural processes and human activities affecting surface and groundwater. Consequently, evaluating the health risks associated with poor groundwater quality necessitates a thorough analysis of water supply

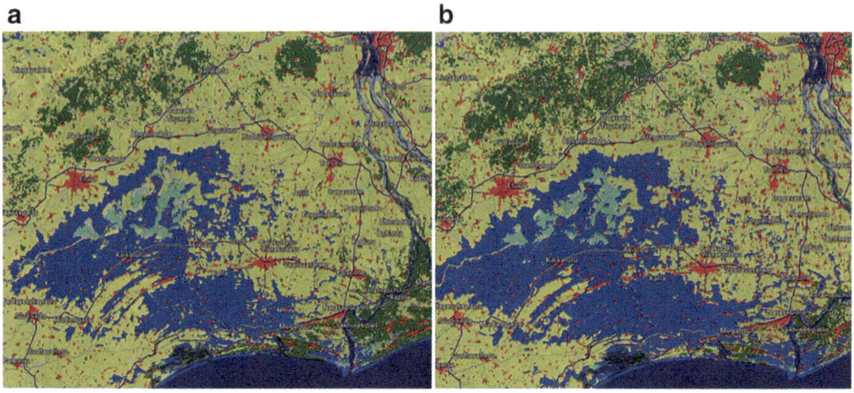

Fig. 3.2 Land use land cover change detection in Godavari Krishna inter-delta in (**a**) 2017 and (**b**) 2023

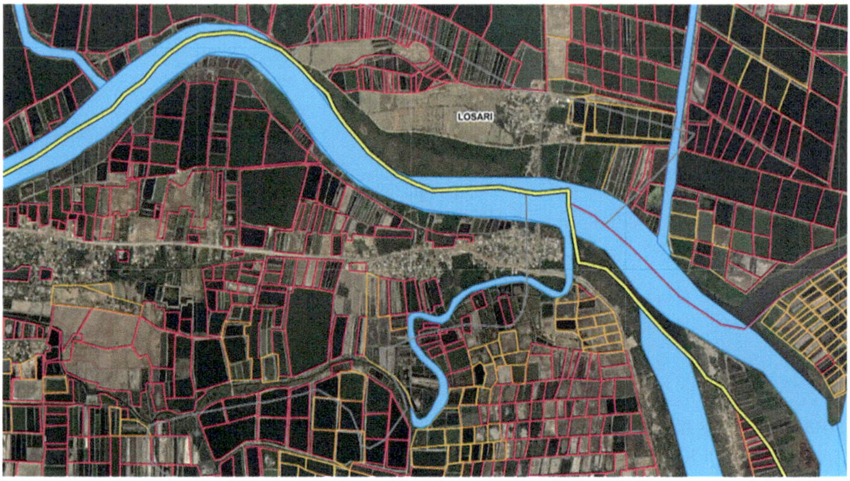

Fig. 3.3 Inland aquaculture ponds along the Upputeru River in Godavari Delta

conditions. However, water quality is strongly influenced by each specific location's unique hydrogeographic and hydrogeochemical characteristics. Figure 3.3 shows the aquaculture ponds along the Upputeru River catchment with Andhra Pradesh State Space Application Center data using remote sensing. In this area, as surface water bodies were contaminated with aquaculture effluents, groundwater is the major source for drinking purposes.

Moreover, groundwater is extensively employed in aquaculture ponds. Rainwater penetrates into the surface of land, percolating through soil and vadose zone. Water is contained within the subsoil voids of the geological structure. Before infiltrating an aquifer, water infiltrates through the vadose zone, potentially becoming deficient in dissolved oxygen and enriched in carbon dioxide as a result of soil organism respiration, particularly in high temperatures. Consequently, groundwater often displays a low oxidation–reduction potential due to oxygen exhaustion, which frequently leads to a relatively low pH caused by increased carbon dioxide content (Boonman et al. 2024).

Extended contact between water and the geological matrix of an aquifer promotes the dissolution of minerals, resulting in groundwater frequently exhibiting higher concentrations of dissolved compounds compared to surface water in a certain region. Minerals have varying solubility; thus, the composition of groundwater can differ significantly between aquifers. On the other hand, in coastal regions, aquifers may contain water from marine sources that has been displaced by freshwater, while the cation exchange sites within the geological matrix of the aquifer remain saturated with sodium (Boonman et al. 2024). When water infiltrating the aquifer traverses limestone, it acquires elevated quantities of calcium and bicarbonate. Similarly, groundwater may exhibit other several water quality issues, including increasing levels of phosphate, nitrate, or ammonia nitrogen, and coloring due to humic compounds. In order to mitigate such problems, it becomes essential to treat acidic water in hatcheries with liming agents.

3.2.3 Challenges Due to Overextraction, Contamination, and Salinity Intrusion in Coastal Regions

Excessive groundwater exploitation, pollution, and salinity intrusion pose substantial issues for coastal areas. A significant concern is the contamination of drinking water, as saltwater intrusion can compromise freshwater sources. This frequently requires expensive measures such as improved water treatment, moving water intakes, or discovering alternate freshwater sources. At the same time, agriculture faces significant risk due to saline intrusion, which can result in enduring salt deposits in the soil, significantly diminishing crop output and jeopardizing food security. On the other end, coastal ecosystems that are recognized for their biodiversity and ecological significance, are exceedingly susceptible to salinity issues and contamination (Sharma and Naik 2024). Correspondingly, mangroves, wetlands, and estuaries are especially vulnerable to alterations in salinity, which can destabilize their fragile equilibrium and endanger the species reliant on them. Categorically, intensive groundwater extraction further aggravates the issue by diminishing the freshwater pressure head in coastal aquifers, decreasing groundwater discharge to the sea, and permitting seawater to encroach in the land, intensifying the salinity dilemma.

On the contrary, another big issue is with saltwater intrusion resulting from sea level rise such as head-controlled or topographically constrained, and flux-controlled or recharging constrained conditions. In the head-controlled intrusion scenario, the inland groundwater level remains constant, while groundwater discharge to the ocean fluctuates. Conversely, in the flux-controlled intrusion scenario, the increase in groundwater level corresponds with the rise in sea level, provided that the unsaturated zone is sufficient to handle the increased groundwater. As a result, the head-controlled system is more susceptible to saltwater intrusion as the terrestrial groundwater level remains unresponsive to the increasing sea level; consequently, the hydraulic gradient between fresh and saltwater becomes uneven, resulting in a salinity intrusion (Nagaraju et al. 2022). The lateral infiltration of saltwater is nearly irreversible within the timeframe that marks the slow movement of groundwater, which requires almost a few decades to expel saltwater.

Several studies have examined the consequences of saline water intrusion in the estuaries of the Godavari Delta. Climate change plays a significant role in sea-level rise due to the thermal expansion of ocean waters caused by higher temperatures and the addition of meltwater from glaciers. This rise in sea level directly influences salinity intrusion in the Godavari delta, mainly because the region, including its coastal areas, averages only about 1 meter above sea level (Das and Swain 2024). Consequently, even minor sea-level changes significantly affect the mixing boundary where freshwater from rivers meets seawater. The extent of saline intrusion depends on the interaction between river discharge, particularly from the Upputeru River, and downstream tidal mixing. An increase in sea level exerts a more substantial backwater effect on river systems, further intensifying saline intrusion. Projections indicate that by 2050, sea levels could rise by approximately 30 cm compared to the 2010 baseline, leading to an estimated 10 km inland expansion of saline intrusion in estuarine areas (Vu et al. 2024). This highlights the increasing vulnerability of the delta to future sea-level changes.

3.3 Impacts of Climate Change on Groundwater Resources in the Delta Ecosystem

Climate change profoundly affects groundwater resources, essential for aquaculture. Severe weather events, including floods, droughts, and storms, change groundwater recharge patterns, and impact its availability for aquaculture activities. These occurrences exacerbate structural damage in coastal and riparian zones, impeding aquaculture infrastructure and services while threatening public health. On the other side, rising sea levels and environmental changes are likely to have a direct impact on soil salinity. This impacts groundwater quality and upsets ecosystems reliant on it, particularly wetlands, which are essential habitats for fish production. Moreover, alterations in river headwaters and diminished glacier melt affect groundwater replenishment, hence influencing aquaculture operations dependent on a stable water supply.

Alterations in water temperature, precipitation patterns, and ocean current dynamics affect the hydrological cycle, resulting in fluctuations in groundwater recharge rates. These alterations jeopardize wetlands, some of the most affluent ecosystems, which rely on groundwater to maintain biodiversity and deliver vital functions for aquaculture (Choudhury et al. 2021). Climate-induced alterations lead to groundwater depletion and contamination, adversely impacting the health of aquatic food chains essential for aquaculture and local economies. Ocean acidification, along with increasing sea surface temperatures, intensifies these issues by disturbing aquatic ecosystems, and shellfish. Coral bleaching incidents and the ensuing deterioration of ecosystems exacerbate the stress on groundwater-dependent environments vital for sustainable fisheries.

The socioeconomic intricacies of climate change, including heightened expenses for water access, alterations in aquaculture methods, and technological requirements, exert strain on groundwater resources. Innovations are necessary to adapt aquaculture to these constraints, ensuring the sustainable utilization of groundwater in the face of changing climatic circumstances. These effects underscore the pressing necessity for cohesive water resource management and climate-adaptive techniques in aquaculture.

3.3.1 Changes in Precipitation Patterns and Recharge Rates

Alterations in precipitation patterns will impact aquaculture productivity and sustainability in two contrasting manners: heightened rainfall (flooding) and intervals of diminished or absent rainfall (drought). Moreover, the hazards associated with drought events are expected to increase at 2 °C of global warming compared to 1.5 °C in a specific region, although patterns of flooding events remain challenging to predict with certainty (Ayugi et al. 2023). Elevated rainfall levels, especially when manifested, will heighten production risks in lowland regions. The dangers encompass the loss of aquatic species from ponds during floods, the invasion of ponds by undesirable species, and damage to ponds due to infilling and erosion of walls. The macro algal productivity, including kelp, may be influenced by increased rainfall, which can provide diverse nutrient loadings to nearshore ecosystems whereas the fluctuations in nutrient loading due to varied precipitation may preferentially benefit invasive short-lived algae species. On the contrary, drought occurrences can induce water stress, resulting in shortages and quality degradation that adversely affect aquaculture productivity and sustainability. Nonetheless, it is essential to further examine how various species and life stages of fish, particularly those of economic significance, may react to alterations in precipitation patterns.

The connection between rainfall patterns and groundwater level and quality across different areas of the Godavari Delta offers valuable insights into ongoing land deformation processes. In this region, groundwater depth varies significantly, ranging from 0.5 to 10 m (Syvitski et al. 2022). This indicates that most delta plains lie within a shallow aquifer zone. This study analyzed groundwater levels measured at eight distinct depths between 2019 and 2024 to assess their influence on land deformation, with Table 3.2 presenting the comparative groundwater level data for 2019 and 2024 that were specifically observed for this research.

Table 3.2 Groundwater level variations in the Godavari Delta region

Location	Average groundwater level in meters	
	2019	2024
Bhimavaram	2.56	3.63
Kallavapudi	7.44	9.73
Malavanatippa	2.25	3.28
Losari	1.54	2.81
Kalipatnam	0.56	1.52
Sessali	6.88	10.45
Kallakuru	3.52	5.42
Akividu	2.21	3.25

In the Godavari Delta region, a previous study by Reshma et al. (2023) reported that various landforms near the river mouth's distributaries characterize the Godavari Delta's coastline (Reshma et al. 2023). Coastal areas such as Kothapallem and Anthirvedi experienced uplift and subsidence during different periods. From 2007 to 2011, Kothapallem showed uplift rates ranging from 0 to +15 mm/year, which increased to +9 to +20 mm/year between 2017 and 2020 (Vamsi Nagaraju and Sunil 2024). In contrast, Anthirvedi exhibited subsidence, with rates of −1 to −11 mm/year from 2007 to 2011 and − 0.6 to −13 mm/year from 2017 to 2020.

3.3.2 Temperature Effects on Groundwater Demand and Quality

Water temperature significantly influences the quality of groundwater and aquaculture by impacting multiple aspects. Elevated temperatures diminish water's capacity to retain oxygen, resulting in dissolved oxygen depletion, and expedite chemical processes such as ionization and mineral dissolution. Fish have accelerated growth in warmer water as a result of heightened metabolism, with an ideal temperature range of 25–32 °C. Increased temperatures facilitate plankton proliferation, which depletes carbon dioxide and bicarbonate, resulting in more significant pH variations between day and night. Moreover, elevated water temperatures decrease viscosity, facilitating the rapid sedimentation of suspended particles. Groundwater subjected to elevated temperatures generally exhibits a reduced pH due to carbon dioxide accumulation and increased concentrations of components such as manganese, with natural groundwater quality commonly affected within a temperature range of 5–20 °C (Cravotta III et al. 2022). Climate change is further projected to elevate global river temperatures by 0.8–1.6 °C by the century's end, thereby affecting aquatic systems (Morovati et al. 2024). To mitigate these issues in aquaculture, it is essential to select species that can withstand local temperature variations and to schedule stocking and harvesting activities to coincide with optimal temperature conditions.

3.4 Geospatial Techniques for Groundwater Assessment

3.4.1 Remote Sensing and GIS Methodologies for Aquaculture Regions

In recent decades, the scientific community has increasingly utilized remote sensing technology to characterize and monitor various systems on both local and global scales. This technology has progressed from solely visual imagery to multispectral photography. Above and beyond, orbital remote sensing data, owing to its comprehensive and multispectral characteristics, offers extensive information over vast and inaccessible regions at regular intervals. This has rendered remote sensing technique as an invaluable instrument in the evaluation of coastal resources. Additionally, the optical satellite sensors, which quantify sunlight reflection in the visible and infrared regions of the electromagnetic spectrum, are predominantly utilized in aquaculture research. Furthermore, these sensors encompass portions of the electromagnetic spectrum, specifically the visible spectrum (blue, green, and red) and near-infrared (VNIR) from 0.4 to 1.3 μm, the mid-infrared (MIR) from 1.55 to 1.75 μm, and the thermal infrared (TIR) from 3.0 to 15.0 μm (Li et al. 2024).

However, the blue band (0.45–0.52 μm) enhances the penetration of water bodies and facilitates the investigation of land use, soil, and vegetation characteristics. The green band (0.52–0.59 μm) is specifically designed to detect the upper reflection of vegetation within the green spectrum, positioned between the blue and red chlorophyll-a absorption bands (Gao et al. 2024). The red band (0.62–0.69 μm) is a crucial channel for distinguishing vegetation. It is also beneficial for delineating soil and geological boundaries. So, all these attributes are even crucial for aquatic resources management. Moreover, terrestrial-oriented satellite instruments, including the Indian RS Satellite (IRS) series, Landsat Thematic Mapper (TM), Enhanced Thematic Mapper Plus (ETM+), and SPOT (Système Pour observation de la Terre), with their high spatial resolution, have been utilized to map shallow bottom features and surrounding land areas by researchers.

3.4.2 Spatial Distribution of Groundwater Dynamics

The inland movement of the fresh-saline water boundary poses significant challenges to water management in the Godavari Delta, as rising salinity compromises groundwater quality and reduces its usability. In this study, 125 samples were analyzed, and 80 exceeded the recommended limits for total dissolved solids. Figure 3.4 shows the groundwater geochemistry in the Godavari delta from the collected samples. Correlation analysis indicated no substantial relationship between geochemical patterns and either depth or proximity to the coast. Elevated electrical conductivity levels in the coastal region were linked to high concentrations of HCO_3^- rather than Cl^- from seawater, suggesting that factors such as ion exchange and organic matter

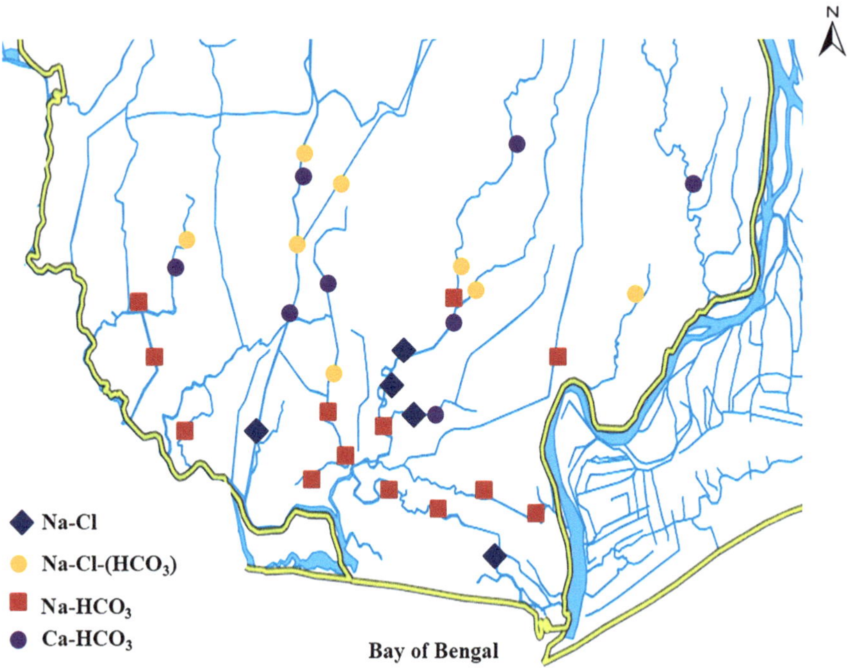

Fig. 3.4 Groundwater geochemistry in aquaculture zone

oxidation contribute to the complexity of groundwater composition. Additionally, soil salts, deltaic clayey sands, coastal clayey sands, aquaculture practices, and marine salts contribute to increased sodium concentrations in groundwater, underlining the intricate interplay of natural and anthropogenic factors affecting salinity in the region.

3.5 Conclusions

Raising awareness and fostering a deeper understanding of groundwater management, particularly in delta regions, is vital for enhancing cross-border cooperation in water resource management. Such efforts are essential to ensuring equitable, efficient, and sustainable water use, thereby supporting economic growth and societal well-being in high-density inland aquaculture zones.

Remote sensing technology has proven invaluable across various agricultural and aquaculture applications. This chapter highlights the role of remote sensing in advancing socio-hydrological research and guiding science-based management practices. It aims to inform policy development for sustainable groundwater management and ecosystem conservation in the Godavari–Krishna inter-delta region. By leveraging remote sensing, this work provides a robust foundation for promoting sustainable ecological and economic development in deltaic environments.

References

Ayugi BO, Chung ES, Zhu H, Ogega OM, Babousmail H, Ongoma V (2023) Projected changes in extreme climate events over Africa under 1.5° C, 2.0° C and 3.0° C global warming levels based on CMIP6 projections. Atmos Res 292:106872

Bashir MA, Xu Y, Wang H, Zhang Y, Sun W, Aon M et al (2025) Integrated rice-crab co-culture system shows capability to reduce greenhouse gases emission and global warming potential. Aquaculture 598:742047

Boonman J, Harpenslager SF, van Dijk G, Smolders AJ, Hefting MM, van de Riet B, van der Velde Y (2024) Redox potential is a robust indicator for decomposition processes in drained agricultural peat soils: a valuable tool in monitoring peatland wetting efforts. Geoderma 441:116728

Chengula AA, Munang'andu HM, Mutoloki S, Evensen Ø (2024) Climate change and viral diseases in aquaculture. In: Aquaculture virology. Academic, pp 109–127

Choudhury M, Sharma A, Singh P, Kumar D (2021) Impact of climate change on wetlands, concerning Son Beel, the largest wetland of North East, India. In: Global climate change. Elsevier, pp 393–414

Cravotta CA III, Senior LA, Conlon MD (2022) Factors affecting groundwater quality used for domestic supply in Marcellus Shale region of North-Central and North-East Pennsylvania, USA. Appl Geochem 137:105149

Dalu T, Chiloane C, Dondofema F, Dube T, Leshaba EN, Masina FM, Munyai LF (2024) Application of remote sensing techniques to monitor climate variability effects on groundwater-dependent ecosystems. In: Remote sensing of climate, pp 133–157

Das A, Swain PK (2024) Navigating the sea level rise: exploring the interplay of climate change, sea level rise, and coastal communities in India. Environ Monit Assess 196(11):1010

Dogeje F, Msuya FE, Ngaruko D, Mohamed AS (2025) Blue economy dynamics: does women's participation in seaweed aquaculture influence their income levels? Empirical evidence from Zanzibar. Aquac Int 33(1):27

Ende S, Henjes J, Spiller M, Elshobary M, Hanelt D, Abomohra A (2024) Recent advances in recirculating aquaculture systems and role of microalgae to close system loop. Bioresour Technol 407:131107

Gao S, Yan K, Liu J, Pu J, Zou D, Qi J et al (2024) Assessment of remote-sensed vegetation indices for estimating forest chlorophyll concentration. Ecol Indic 162:112001

Gottumukkala SB, Thotakura VN, Gvr SR, Chinta DP, Park R (2024) Balancing aquaculture and estuarine ecosystems: machine learning–based water quality indices for effective management. Environ Sci Pollut Res:1–17

Ha QK, Phung TH, Phan NL, Minderhoud PS, Le Vo P, Essink GO (2024) Groundwater resource challenges and abstraction-induced land subsidence in the Vietnamese Mekong Delta. In: The Mekong River Basin, pp 421–451

Ibrahim A, Wayayok A, Shafri HZM, Toridi NM (2024) Remote sensing technologies for modelling groundwater storage dynamics: comprehensive review. J Hydrol X 23:100175

Jaikumar M, Ramadoss D, Sreekanth GB, Smrithi K, Parihar RD (2023) Regional impacts of COVID-19 pandemic on aquaculture and small-scale fisheries: Insights and recovery strategies in India. Aquaculture 570:739403

Jiang Z, Hu S, Deng H, Wang N, Zhang F, Wang L et al (2024) Detection and automatic identification of loess sinkholes from the perspective of LiDAR point clouds and deep learning algorithm. Geomorphology 465:109404

John A, Kaur A, Bhat KA, Ganie SA, Nugroho C, Handoko I, Wani AK (2024) Adapting to climate extremes: implications for insect populations and sustainable solutions. J Nat Conserv 126602:126602

Kim MJ, Han IS, Lee JS, Kim DH (2024) A climate change risk assessment in aquaculture in Korea. Clim Chang 177(6):93

Kumar R, Singh CK, Misra S, Singh BP, Bhardwaj AK, Chandra KK (2024) Water biodiversity: ecosystem services, threats, and conservation. In: Biodiversity and bioeconomy. Elsevier, pp 347–380

Li J, Wong MS, Shi G, Nichol JE, Lee KH, Chan PW (2024) Advanced algorithms on monitoring diurnal variations in dust aerosol properties using geostationary satellite imagery. Remote Sens Environ 303:113996

Miyagi T, Ikeda K, Ishikawa H, Doan L, Thanh NK, Van Tien P et al (2024) Interpretation and mapping for the prediction of sites at risk of landslide disasters: from aerial photography to detection by DTMs. In: Progress in landslide research and technology, vol 3 Issue 1. Springer, Cham, pp 15–61

Morovati K, Tian F, Pokhrel Y, Someth P, Shi L, Zhang K et al (2024) Fishery and agriculture amidst human activities and climate change in the Mekong River: a review of gaps in data and effective approaches towards sustainable development. J Hydrol 644:132043

Nagaraju TV, Malegole SB, Chaudhary B, Ravindran G (2022) Assessment of environmental impact of aquaculture ponds in the western delta region of Andhra Pradesh. Sustain For 14(20):13035

Nagaraju TV, Sunil BM, Chaudhary B, Prasad CD, Gobinath R (2023) Prediction of ammonia contaminants in the aquaculture ponds using soft computing coupled with wavelet analysis. Environ Pollut 331:121924

Nagaraju TV, Bala GS, Bonthu S, Mantena S (2024a) Modelling biochemical oxygen demand in a large inland aquaculture zone of India: implications and insights. Sci Total Environ 906:167386

Nagaraju TV, Malegole SB, Chaudhary B, Ravindran G, Chitturi P, Chinta DP (2024b) Novel assessment tools for inland aquaculture in the western Godavari delta region of Andhra Pradesh. Environ Sci Pollut Res 31(25):36275–36290

Naik NA (2024) Livelihood perspectives and socioeconomic implications of technological transitions on the small-scale fisheries: an analytical study of Ramponkar Communities in Goa (Doctoral dissertation, Kerala University of Fisheries and Ocean Studies)

Pandey A, Pathan MA, Ananthan PS, Sudhagar A, Krishnani KK, Sreedharan K et al (2024) Stocking for sustainable aqua-venture: optimal growth, yield and economic analysis of Penaeus vannamei culture in inland saline water (ISW) of India. Environ Dev Sustain 26(3):6913–6942

Pradhan SK, Nayak PK, Haque CE (2023) Mapping social-ecological-oriented dried fish value chain: evidence from coastal communities of Odisha and West Bengal in India. Coasts 3(1):45–73

Rana KK, Pachu AV, Jeeva V, Rao NR, Sekhar A, Singh AN, Kumar S (2024) Enhancing sustainability: reclamation and rehabilitation strategies for restoring mined-out lands in India to mitigate climate change impacts. In: Forests and climate change: biological perspectives on impact, adaptation, and mitigation strategies. Springer, Singapore, pp 573–603

Reshma KN, Agrawal R, Ramakrishnan R, Sreejith KM, Rajawat AS (2023) Land subsidence studies in the Godavari Delta regions of the East coast of India using ALOS and Sentinel 1 data. Eco Inform 78:102373

Roy A, Kumar S, Rahaman M (2024) Exploring climate change impacts on rural livelihoods and adaptation strategies: Reflections from marginalized communities in India. Environ Dev 49:100937

Saizen I, Funakawa S, Okamoto Y, Hoang TC, Ngo BH, Tran NNK et al (2024) Livelihood and its surrounding environments in rural areas. In: Livelihood and the environment in Vietnam. Springer, Singapore, pp 49–172

Sharma LK, Naik R (2024) Conservation of saline wetland ecosystems: an initiative towards UN decade of ecological restoration. Springer

Syvitski J, Anthony E, Saito Y, Zăinescu F, Day J, Bhattacharya JP, Giosan L (2022) Large deltas, small deltas: toward a more rigorous understanding of coastal marine deltas. Glob Planet Chang 218:103958

Tchonkouang RD, Onyeaka H, Nkoutchou H (2024) Assessing the vulnerability of food supply chains to climate change-induced disruptions. Sci Total Environ 920:171047

Vamsi Nagaraju T, Sunil BM (2024) Briefing: Intensive inland aquaculture ponds: challenges and research opportunities. Environ Geotech 40(XXXX):1–5
Vu MT, Luu C, Bui DQ, Vu QH, Pham MQ (2024) Simulation of hydrodynamic changes and salinity intrusion in the lower Vietnamese Mekong Delta under climate change-induced sea level rise and upstream river discharge. Reg Stud Mar Sci 78:103749
Zhan C, Dai Z, Yin S, Carroll KC, Soltanian MR (2024) Conceptualizing future groundwater models through a ternary framework of multisource data, human expertise, and machine intelligence. Water Res 257:121679

Chapter 4
Harnessing Machine Learning and Deep Learning for Water Quality Analysis in Deltaic Environments

G. Sri Bala, T. Vamsi Nagaraju, and Ch. Durga Prasad

Abstract Predicting water quality is crucial for managing water resources and preventing pollution. Predictive analytics research for different water bodies in delta environments has not kept up with environmental demands. This chapter fills this gap by doing a thorough study and predicting water quality measurement data from numerous dispersed water bodies, particularly inside the delta regions. A case study in South India's delta region was also examined to forecast the aquaculture zone's water quality utilizing machine learning and deep learning methodologies. Ammonia levels in aquaculture water bodies were predicted using random forest (RF), support vector machines (SVM), convolutional neural networks (CNN), and forward neural networks (FNN). The study's empirical findings demonstrate the system's exceptional forecasting capabilities. This chapter provides particular recommendations for water quality monitoring, treatment, and management strategies based on these findings, which are adapted to the unique requirements of the aquaculture zone in the delta regions. These contributions can potentially help environmental managers and policymakers make better decisions.

Keywords Delta region · Inland aquaculture · Water quality · Machine learning

G. S. Bala · T. V. Nagaraju (✉)
Department of Civil Engineering, SRKR Engineering College, Bhimavaram, Andhra Pradesh, India

Centre for Sustainable Inland Aquaculture, SRKR Engineering College, Bhimavaram, India

C. D. Prasad
Department of Electrical and Electronics Engineering, SRKR Engineering College, Bhimavaram, India

T. V. Nagaraju, G. S. Bala (eds.), *Delta Ecosystems*, Environmental Contamination Remediation and Management, https://doi.org/10.1007/978-3-031-97920-0_4

4.1 Introduction

The dense canal networks in the deltas are seen as a strength for economic growth and foster an environment conducive to other social activities (Lewis and Ernstson 2019). However, water resources in the delta ecosystem have significantly been impacted in recent years by the effects of climate change and operations on the mainstream of any delta in the world (Dang et al. 2018). Upstream, the river significantly impacts the delta's water resources. As a result, downstream regions have faced numerous challenges due to the effects on the mainstream. Expanding the upstream hydroelectric dam's reservoirs may lessen sediment loads, worsen water shortages, increase erosion, and cause a rise in relative sea level (Kondolf et al. 2018). Variations in river flow also have an impact on the region's agricultural output. This water resource has also degraded due to a significant increase in human needs and the waste released from anthropogenic activities like industry, agriculture, and aquaculture (Vamsi Nagaraju and Sunil 2024). Rivers are mostly used to discharge wastewater from homes, businesses, and farms; nevertheless, they are also susceptible to contamination (Singh et al. 2022). As a result, maintaining surface water quality has become crucial to ensuring sustainable growth. To manage water resources, baseline and activity-affected areas must be monitored (Liu et al. 2013). This quickly mitigates adverse effects and offers insights into geographical variations in surface water quality. This monitoring benefits delta regions since it may impact the downstream regions' water quality (Nagaraju et al. 2024a).

Point discharges significantly impact surface water quality in addition to diffuse inputs. Because their untreated effluent contains significant levels of suspended particles, organic matter, nitrogen, phosphorus, and bacteria, the aquaculture extraction sectors are frequently to blame for surface water pollution in the deltas (Cochard 2017). Centralized wastewater treatment facilities are not present in every industrial park and export processing zone that was operationalized within the project area. Therefore, because of heavy pollution and stagnation, areas with high population densities and significant industrial activity suffer from a severe deterioration in water quality (Madhav et al. 2020).

Despite their many benefits, delta systems are therefore highly vulnerable to climate change-related events (such as more frequent flooding and more prolonged droughts, acid sulfate soils, acid water flows, and a lack of freshwater supplies), as well as extreme pollution from industrial, agricultural, and aquaculture operations and emissions from both urban and rural areas (Birkmann et al. 2012). Land use and surface water quality are strongly correlated, according to studies conducted in the upper portions of the deltas (Zhao et al. 2015). The type of land use significantly influences understanding and preventing the deterioration of water quality. Though it has not been investigated in a large-scale freshwater area in this region, the impact of various land use types on surface water could be crucial to future sustainable land use and water planning (Nagaraju et al. 2024b). Investigations like this can benefit significantly from information-based support technologies like geographic information systems. Both a comprehensive approach to analyzing local conditions and a comprehensive assessment of the water quality of large regions are feasible. As a result, remote sensing has emerged as a significant and effective tool for regional land use and water planning (Mashala et al. 2023). Additionally, using statistical techniques, like principal component analysis, enables a more thorough examination of the variance of the measured data to pinpoint the sources of pollution and identify the important water pollution factors. According to recent studies, principal component analysis is valuable for examining water quality measurement data and enhancing the monitoring system (Nagaraju et al. 2023).

Many studies and applications of machine-learning-based technology for water quality prediction in the delta region have been conducted in recent years. To create precise prediction models and facilitate early warning, machine learning uses significant historical water quality data in the context of water quality prediction. This technology has several benefits (Sun and Scanlon 2019). First, it makes it easier to monitor and forecast water quality continuously and in real time, improving the efficacy and responsiveness of water quality management (Mohanty et al. 2024). Second, more accurate forecasts can be produced using machine learning algorithms that automatically determine and adjust to the intricate relationships in water quality data (Bui et al. 2020).

Furthermore, by incorporating meteorological data and additional environmental parameters, these models can enhance the precision and dependability of water quality forecasting (Yaseen et al. 2019). However, estimating the water quality of the delta region presents additional difficulties and constraints for machine learning techniques. Problems like missing data and poor data quality may impact the model's performance. Furthermore, a certain amount of professional expertise and experience is needed for training and parameter selection. This chapter examines a case study of the South Indian delta region and highlights the value of machine learning models in water quality monitoring there.

4.2 Machine Learning for Water Quality Monitoring in Delta Ecosystems

Numerous researchers have applied machine learning to address issues in water treatment and handling systems, such as real-time monitoring, prediction, tracking the sources of pollutants, estimating their concentrations, allocating water resources, and optimizing water treatment equipment (Alprol et al. 2024; Drogkoula et al. 2023). Applications in surface water human-generated wastewater from both industries and municipalities is now the primary cause of declining urban water quality. Machine learning in studies on surface water quality has gained popularity (Huang et al. 2021). Much work has gone into making machine learning models better and increasing the accuracy of their predictions (Gottumukkala et al. 2024). Gathering data is one of the most important steps in creating machine learning models. Water system management can use integrated and regular water quality monitoring results as benchmarks. Environmental agencies commonly use traditional environmental monitoring techniques (Zhang et al. 2017). However, practical challenges for real-time tracking limit the use of traditional approaches.

Prediction accuracy in machine learning applications is typically correlated with the quality of the learning dataset and the model selection. Support vector machines (SVMs) and artificial neural networks (ANNs) have demonstrated exceptional performance in forecasting water quality components (Wang et al. 2022). In certain situations, SVM may outperform ANN in terms of prediction accuracy and generalization capacity. One explanation is that nonlinear disturbances significantly impact the accuracy of the ANN because the neural network's model parameter optimization is unstable (Shi et al. 2020). Furthermore, SVM is more successful than ANN at minimizing the generalization error because it employs an upper bound on the error rather than lowering the training error. Because river systems undergo intricate, dynamic changes over time, the best strategy to manage delta zones is to either monitor water quality in real time or, in the event that monitoring conditions are not suitable, create predictions based on other data. Long short-term memory (LSTM) networks and other algorithms can handle nonseasonal and changing time-series data; researchers have validated their effectiveness for modeling water-quality data.

4.3 Case Study of Godavari Delta Inland Aquaculture Water Quality Monitoring

4.3.1 Data Collection and Preprocessing

In this study, between January 2023 and December 2023, a total of 230 samples were tested for various parameters, and a dataset was generated to predict ammonia levels in aquaculture water. These datasets include nine input variables: calcium (Ca), magnesium (Mg), hardness, alkalinity, pH, carbonates, bicarbonates, dissolved oxygen (DO), and ammonia, which serves as the output variable. The water samples were taken from various earthen aquaculture shrimp ponds, and all the water quality parameters were measured in the laboratory according to American Public Health Association (APHA) standards. The statistical properties of these parameters are presented in Table 4.1. From the dataset, low DO and high ammonia levels were observed due to poor maintenance of water quality and irregular monitoring in the shrimp ponds. In the earthen shrimp ponds, temperature plays a significant role in regulating a variety of biological and physicochemical processes. The temperature at each study's sampling site was roughly the same (28–30 °C), within the 24–34 °C permissible range for aquaculture.

Various evaluation metrics are essential in assessing prediction models' effectiveness. In the case of ammonia prediction, four metrics, R^2, root mean square error (*RMSE*), mean square error (MSE), and mean absolute error (*MAE*), were used to analyze the model efficiency.

Table 4.1 Statistical data on water parameters in shrimp ponds

Parameter	pH	Salinity	Alkalinity	Carbonates	Bicarbonates	DO	Ca	Mg	Hardness	Ammonia
Minimum	6.83	1.10	70.00	0.00	21.00	0.80	53	65.00	201.00	0.00
Maximum	9.42	11.20	640.00	110.00	640.00	6.00	1610.00	1520.00	3860.00	1.86
Mean	8.27	5.30	200.00	20.00	180.00	4.70	214.50	272.50	1535.00	0.19
Standard deviation	0.35	1.68	82.90	16.28	80.02	0.87	138.62	135.36	555.30	0.34

4.3.2 Ammonia Prediction in Earthen Shrimp Ponds Within Delta Ecosystems

Predicting ammonia levels in the aqua ponds is essential to ensure a healthy ecosystem, and estimating the amount of ammonia using other water quality indicators such as pH, salinity, alkalinity, carbonates, bicarbonates, DO, Ca, Mg, and hardness is effective because these parameters are interconnected with complex relationships. Understanding these relationships allows for more accurate predictions of ammonia levels, enabling better management and control of water quality, thereby ensuring a healthier environment for aquatic life and safe water for human use. Therefore, soft computing tools are suggested to predict the ammonia rather than traditional statistical methods. Furthermore, these techniques can be integrated into real-time monitoring systems, providing instant predictions of ammonia levels to support timely decision-making and immediate responses to potential water quality issues. In this work, random forest (RF), SVR, convolutional neural network (CNN), and forward neural network (FNN) are used to predict ammonia with other water indicators. The performance of these techniques is evaluated with the help of root mean square error (RMSE), MSE, MAE, and R^2.

RF is a decision tree-based ensemble learning technique. CNN is a deep learning approach frequently used to recognize picture tasks; it is adapted for forecasting ammonia in aquaculture waters. SVM is an algorithm for supervised learning utilized in regression and classification analyses. Comparative research is done to assess each model's performance. The models generated using machine learning and deep learning methods are evaluated and contrasted according to how well they predict ammonia. Using the ideas of bagging and random feature selection, the RF algorithm is a potent ensemble ML technique that creates a robust prediction model. In ensemble approaches, the predictive performance is enhanced by merging several independent models. By taking into account only a random subset of features at each split point when building trees, RF adds another degree of unpredictability to the process. As a result, the model is more resilient to noise and outliers, and the trees are better diversified. Figure 4.1 shows the ammonia levels that were predicted using machine learning.

The supervised learning method, SVR, applied to regression tasks, aims to maximize the margin between data points while identifying the optimal line, or hyperplane, that fits the information points. According to the definition of "obstinate generalization ability," the SVR model performs well and resists overfitting. Put differently, its ability to efficiently extrapolate its forecasts to novel and unexplored datasets renders it a dependable predictor. The stated R^2 value of 0.924 indicates the SVR ammonia prediction model's fit quality.

The superiority of these machine and deep learning models is strongly agreed with the performance metrics reported in Table 4.2. The experimental data in this instance exhibits a strong correlation with the SVR model, indicating that the model well represents the fundamental characteristics and tendencies in the ammonia of aquaculture waters. The potential to transform the management of inland aquaculture water is why the application of CNN for ammonia prediction is so important.

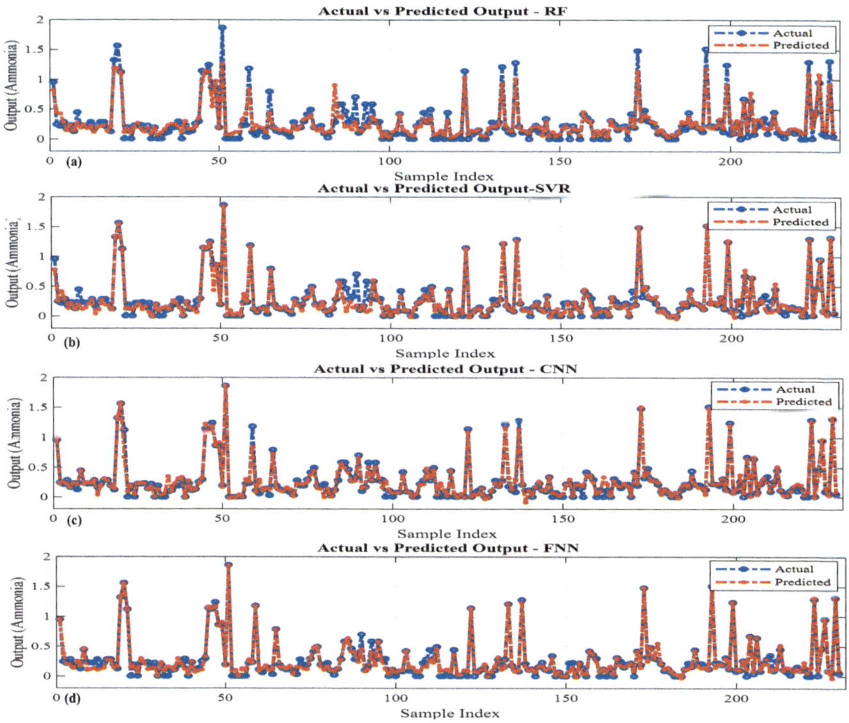

Fig. 4.1 Ammonia prediction plots using various techniques: (**a**) RF, (**b**) SVR, (**c**) CNN, and (**d**) FNN

Table 4.2 Ammonia prediction models' performance

Method	RMSE	MSE	MAE	R^2
RF	0.1262	0.0159	0.0851	0.8613
SVR	0.0932	0.0087	0.0516	0.9245
CNN	0.0760	0.0058	0.0467	0.9497
FNN	0.0701	0.0049	0.0501	0.9572

CNN algorithm works well for tasks involving pattern and picture recognition. It can accurately analyze the characteristics and correlation of aquaculture waters in this context to estimate the ammonia levels in the water.

4.4 Conclusions

In light of urbanization and the depletion of clean water supplies, water resource management has grown significantly. In addition to advanced approaches for modeling and predicting water quality, machine learning and deep learning development have produced effective ways to handle the complex data gathered from many water sources to assess and forecast trends and variances in delta ecosystems. This chapter introduced the many learning techniques in deep and machine learning and the steps necessary for accurate water quality modeling and prediction. The Godavari Delta's intense aquaculture zone uses a variety of machine learning and deep learning algorithms to estimate the water quality. According to experimental studies, each model can forecast ammonia to a different extent. However, no single model is better, as superiority varies depending on the studied issue. For some problems, one paradigm may work well, but not for others. This makes the application of machine learning and deep learning in determining water quality more difficult because of the complexity of specificity, as models might vary greatly based on the water body and factors.

In contrast, these models could be safely used as a productive replacement for models that are based on physical data. Beyond their ability to track and evaluate water quality, these techniques help address environmental issues by spotting potential hazards in the future. Therefore, the best action is to create a more consistent and adaptable predictive model that works well over various factors and sources. In order to manage delta ecosystems more effectively, future developments in machine learning and deep learning as a water quality tool may also concentrate on assessing and forecasting patterns in the fluctuations of various organic and inorganic contaminants in diverse water bodies.

References

Alprol AE, Mansour AT, Ibrahim MEED, Ashour M (2024) Artificial intelligence technologies revolutionizing wastewater treatment: current trends and future prospective. Water 16(2):314

Birkmann J, Garschagen M, Van Tuan V, Binh NT (2012) Vulnerability, coping and adaptation to water related hazards in the Vietnamese Mekong Delta. In: The Mekong Delta system: interdisciplinary analyses of a river delta, pp 245–289

Bui DT, Khosravi K, Tiefenbacher J, Nguyen H, Kazakis N (2020) Improving prediction of water quality indices using novel hybrid machine-learning algorithms. Sci Total Environ 721:137612

Cochard R (2017) Coastal water pollution and its potential mitigation by vegetated wetlands: an overview of issues in Southeast Asia. Redef Divers Dynam Nat Resour Manag Asia 1:189–230

Dang TD, Cochrane TA, Arias ME (2018) Future hydrological alterations in the Mekong Delta under the impact of water resources development, land subsidence and sea level rise. J Hydrol Reg Stud 15:119–133

Drogkoula M, Kokkinos K, Samaras N (2023) A comprehensive survey of machine learning methodologies with emphasis in water resources management. Appl Sci 13(22):12147

Gottumukkala SB, Thotakura VN, Gvr SR, Chinta DP, Park R (2024) Balancing aquaculture and estuarine ecosystems: machine learning–based water quality indices for effective management. Environ Sci Pollut Res 31:1–17

Huang R, Ma C, Ma J, Huangfu X, He Q (2021) Machine learning in natural and engineered water systems. Water Res 205:117666

Kondolf GM, Schmitt RJ, Carling P, Darby S, Arias M, Bizzi S et al (2018) Changing sediment budget of the Mekong: cumulative threats and management strategies for a large river basin. Sci Total Environ 625:114–134

Lewis JA, Ernstson H (2019) Contesting the coast: ecosystems as infrastructure in the Mississippi River Delta. Prog Plan 129:1–30

Liu Y, Zhang J, Wang G, Liu J, He R, Wang H et al (2013) Assessing the effect of climate natural variability in water resources evaluation impacted by climate change. Hydrol Process 27(7):1061–1071

Madhav S, Ahamad A, Singh AK, Kushawaha J, Chauhan JS, Sharma S, Singh P (2020) Water pollutants: sources and impact on the environment and human health. In: Sensors in water pollutants monitoring: role of material, pp 43–62

Mashala MJ, Dube T, Mudereri BT, Ayisi KK, Ramudzuli MR (2023) A systematic review on advancements in remote sensing for assessing and monitoring land use and land cover changes impacts on surface water resources in semi-arid tropical environments. Remote Sens 15(16):3926

Mohanty A, Mohanty SK, Mohapatra AG (2024) Real-time monitoring and fault detection in AI-enhanced wastewater treatment systems. In: The AI cleanse: transforming wastewater treatment through artificial intelligence: harnessing data-driven solutions. Springer, Cham, pp 165–199

Nagaraju TV, Sunil BM, Chaudhary B, Prasad CD, Gobinath R (2023) Prediction of ammonia contaminants in the aquaculture ponds using soft computing coupled with wavelet analysis. Environ Pollut 331:121924

Nagaraju TV, Malegole SB, Chaudhary B, Ravindran G, Chitturi P, Chinta DP (2024a) Novel assessment tools for inland aquaculture in the western Godavari delta region of Andhra Pradesh. Environ Sci Pollut Res 31(25):36275–36290

Nagaraju TV, Bala GS, Bonthu S, Mantena S (2024b) Modelling biochemical oxygen demand in a large inland aquaculture zone of India: implications and insights. Sci Total Environ 906:167386

Shi Z, Yao W, Zeng L, Wen J, Fang J, Ai X, Wen J (2020) Convolutional neural network-based power system transient stability assessment and instability mode prediction. Appl Energy 263:114586

Singh N, Poonia T, Siwal SS, Srivastav AL, Sharma HK, Mittal SK (2022) Challenges of water contamination in urban areas. In: Current directions in water scarcity research, vol 6. Elsevier, pp 173–202

Sun AY, Scanlon BR (2019) How can Big Data and machine learning benefit environment and water management: a survey of methods, applications, and future directions. Environ Res Lett 14(7):073001

Vamsi Nagaraju T, Sunil BM (2024) Briefing: intensive inland aquaculture ponds: challenges and research opportunities. Environ Geotech 40(XXXX):1–5

Wang G, Jia QS, Zhou M, Bi J, Qiao J, Abusorrah A (2022) Artificial neural networks for water quality soft-sensing in wastewater treatment: a review. Artif Intell Rev 55(1):565–587

Yaseen ZM, Sulaiman SO, Deo RC, Chau KW (2019) An enhanced extreme learning machine model for river flow forecasting: state-of-the-art, practical applications in water resource engineering area and future research direction. J Hydrol 569:387–408

Zhang Y, Chu C, Li T, Xu S, Liu L, Ju M (2017) A water quality management strategy for regionally protected water through health risk assessment and spatial distribution of heavy metal pollution in 3 marine reserves. Sci Total Environ 599:721–731

Zhao J, Lin L, Yang K, Liu Q, Qian G (2015) Influences of land use on water quality in a reticular river network area: a case study in Shanghai, China. Landsc Urban Plan 137:20–29

Chapter 5
Water Quality Management and Monitoring in Delta Ecosystems

Pragati Singh, Nimai Das, Neetu Arya, and Prem Shankar

Abstract Water quality in delta ecosystems is shaped by factors like climate change, which raises temperatures and triggers harmful algal blooms, and anthropogenic activities such as agriculture and pollution that introduce heavy metals and nutrients. Additionally, natural disasters, geological factors, and tipping points related to sea level rise and flooding further contribute to water degradation. Traditional methods lack the temporal and spatial coverage required for effective management, making remote sensing (RS) technologies increasingly valuable for retrieving water quality parameters (WQPs) in real time. Recent advancements, for example, optical and hyperspectral sensors and artificial intelligence (AI), have improved the accuracy of quality of water measurement by enabling diverse retrieval methodologies. These include empirical, analytical, semi-empirical, and machine-learning algorithms, all of which leverage various data sources like satellites and in situ sensors. Automation in water quality systems, incorporating continuous data acquisition and transmission, facilitates near real-time monitoring, generating big data for predictive modeling. This is particularly important in river deltas, where climate change and human activity intensify water quality issues. RS applications offer decision-makers powerful tools for spatiotemporal WQP analysis, supporting the development of successful guidelines and management procedures. The

P. Singh
Department of Environment Management, Indian Institute of Forest Management, Bhopal, India

N. Das
Department of Environment and Developmental Economics, Indian Institute of Forest Management, Bhopal, India

N. Arya
Department of Climate Change & Forest Influence, Indian Institute of Forest Management, Bhopal, India

P. Shankar (✉)
Department of Forestry & Ecology, Indian Institute of Remote Sensing (ISRO), Dehradun, India

T. V. Nagaraju, G. S. Bala (eds.), *Delta Ecosystems*, Environmental Contamination Remediation and Management, https://doi.org/10.1007/978-3-031-97920-0_5

combination of advanced RS technologies along with traditional methods is crucial for addressing water quality challenges and fostering sustainable river delta ecosystem management in the face of environmental change.

Keywords Water quality · Remote sensing · Machine learning · Delta ecosystem

5.1 Introduction

Coastlines serve as critical interfaces between land, ocean, and inland waters (Kontopoulos et al. 2022). Coastal and riparian zones host diverse ecosystems and have rich biodiversity, making them invaluable for economic, cultural, and environmental reasons (Feyen et al. 2020). Coastal deltas of rivers are low-lying, sediment-shaped landforms that depend on continuous sediment supply to maintain and stabilize the coastline. These deltas are essential for the yield of surface water, carbon sequestration, retention of soil, and preservation of the ecosystem. They offer vital ecosystem services, including natural flood mitigation, water purification, pollutant absorption, and decomposition, which enhance the water quality and support biodiversity (Peng et al. 2023; Wu et al. 2022).

With abundant resources, favorable natural conditions, and strategic geographical locations, deltas have become hubs of economic activity and human development (Anthony 2015). These areas have a major influence on worldwide sustainable advancement. Despite occupying less than 0.5% of Earth's area of land, deltas are host to over 5% of the population of the world. Urban areas within some African and Asian deltas have population densities exceeding 10,000 people per square kilometer. Rapid urbanization in these regions underscores the importance of guiding sustainable development to address the growing pressure on vulnerable environments (Scown et al. 2023). River deltas are often densely populated because of the livelihood opportunities they provide, particularly through agriculture and aquaculture, as well as their role in economic development and trade (Román et al. 2022). For instance, some of the largest urban development networks in China are found in the lowlands that surround the deltas. The Mississippi Delta is North America's main migratory bird flyway and is home to the biggest fishery in the country. The Ciliwung Delta is where Indonesia's capital, Jakarta, was constructed. Among the totally focused deltas are the Ayeyarwady, Ganges, Brahmaputra, Meghna, and Mekong, while Egypt's Nile Delta makes a substantial economic contribution to the country through industry, services, tourism, and agriculture (Loucks 2019).

Although river deltas offer significant opportunities for development, they are increasingly threatened by both human activities and environmental factors (Renaud et al. 2016). Urbanization in these regions often brings challenges, such as Sinking of land, surface sealing, drainage flow control, pollution of the atmosphere, the ground, and water, overexploitation of resources from nature, and substantial disruptions to natural delta systems. In many deltas, agricultural intensity has increased, exacerbates water and soil pollution, reduces biodiversity by altering nutrient and trace element flow, and contributes to land subsidence through excessive groundwater extraction. Large-scale upstream activities, including urban development, industrial and agricultural water withdrawals, and construction of hydropower dams, can severely impact downstream deltas. These effects are compounded by infrastructure development within the deltas, such as distributary channel regulation, irrigation, and salinity management. Such interventions intercept nearly 40% of the world's river discharge and trap approximately one-third of the sediment flux from continents to coastal zones, demonstrating how human engineering increasingly dictates the development of numerous deltas.

Variations in the climate increase this pressure more intensely. Rising sea levels have increased coastal erosion, flooding, and salty water seeping into rivers, canals, aquifers, and soils. Hazards to the environment such as typhoons, storm surges, and extreme tides, along with local shifts in rainfall patterns, intensities, and rising temperatures, exacerbate the vulnerability of these regions. Additionally, human-induced "technical" hazards, such as spillage of oil, chemical incidents, dyke failures, and cracks in levees, impose further strain on delta ecosystems (Renaud et al. 2013). Saline water intrusion (SWI) and salinity patterns in deltas are a consequence of the interplay among morphology, topography, river mouth tidal dynamics, and variations in freshwater and marine flow. Over time, the SWI can significantly degrade water quality, affecting its suitability for drinking, industrial use, and agriculture. Owing to the higher density of seawater, it can intrude beneath freshwater, forming a saltwater wedge that influences the manner in which biological and chemical features are distributed, thereby impacting the overall water quality. Seawater can infiltrate surface water systems and spread inland through rivers, natural channels, or artificial waterways, and it can also permeate groundwater (Reljić et al. 2023). Figure 5.1 shows the drivers of degradation and its impact on the river delta ecosystem.

In response to these challenges, various political and environmental organizations, including the European Commission and the United Nations (UN), have issued guidelines and initiatives to protect vulnerable coastal areas. Notable efforts include Goals 13 and 14 of the United Nations sustainable development agenda, the European Green Deal, and the European Directive for Maritime Spatial Planning, which emphasize climate action and marine conservation. Additionally, the Copernicus Land Monitoring Service introduced the Coastal Zone dataset to support coastal management. Specifically, the 14th Sustainable Development Goal of the UN underscores the relevance of precise shoreline information for protecting the coast (Kontopoulos et al. 2022). Recent research has also addressed environmental risks in coastal zones. Nguyen Kim Hoang and Dinh Tuan Hai, for instance, investigated salinity intrusion in Rach Gia City in 2024, which is located in the Mekong Delta of Vietnam. Their findings provide a comprehensive analysis of salinity intrusion, identify key contributing factors, and offer critical insights for decision-makers. These insights support effective water resource management, mitigation of economic losses, and sustainable land-use planning, thereby protecting essential resources for local communities. Liu et al. (2023) investigated the spatiotemporal dynamics of land use and land cover in the Mekong Delta. Their findings indicated

Fig. 5.1 Drivers of degradation and its impact on the river delta ecosystem

an increase in aquaculture and residential land, coupled with a decline in mangroves, agricultural land, forests, wasteland, and unused land. Management of land and commercial growth initiatives were the main forces behind these changes. Zhou et al. (Wu et al. 2024) explored the impact of urbanization and changing climates on the risk of flooding in the delta of the Ganges, Brahmaputra, and Meghna (GBM). According to their analysis, the risk of flooding would significantly increase as a result of changes in the climate by the end of the significantly contributing to urbanization and coastal flood hazards, which increase the probability of fluvial and pluvial flooding. These findings underscore that adaptive urban planning is essential to minimizing future flood risks and promoting balanced urban growth. Bhadra et al. (2020) assessed groundwater quality in the southern Ganges–Brahmaputra Delta, which is in the Sundarban Biosphere Reserve. Their analysis revealed that the region's groundwater exhibits $NaHCO_3$, NaCl, and mixed-type hydrogeochemical facies with a high Cl/HCO_3 ratio, rendering it unsuitable for drinking because of the entrance of saline water.

The growing susceptibility of river deltas has attracted global attention in recent years. Initiatives that have placed a strong emphasis on scientific cooperation, cooperative applied research, and information sharing include the Delta Coalition, Delta Alliance, Connecting Delta Cities Network, and World Estuary Alliance. These were set up in the month of March 2015 at the third UN Conference on reducing disaster risks in Sendai, Japan. These efforts aim to achieve a thorough knowledge of the changes in the environment that affect the large river delta regions. In order to manage river deltas sustainably, a comprehensive approach is required to assess past and current shifts in the patterns of settlement, land cover and use, ecosystem development, geomorphological changes in rivers and coastlines, and water quality. Achieving this is challenging using traditional land-based surveying methods alone (Kuenzer et al. 2019). This chapter explores the potential of both traditional and advanced technologies for managing and monitoring water quality in delta ecosystems and also emphasizing comparison among them.

5.2 Technology Advancement in Water Quality Management

Having quality data and knowledge is crucial for putting adaptive techniques into practice, enhancing the resilience of river deltas to common threats, and making informed decisions regarding integrated land and water resource management. The possible risks caused by adverse conditions may be decreased through comprehensive and improved monitoring and control of water resources. One important strategy is the quality of water monitoring for controlling pollutants and pathogens that may harm the environment and human health. Traditional analysis of water quality often involves lengthy and labor-intensive fieldwork, requiring water samples to be transported to laboratories for further analysis. This process is not only physically demanding but also poses significant challenges, including exposure to dangerous contaminants like harmful algal blooms (HABs), which can create secondary risks for field personnel. Technological advances have transformed water quality monitoring methods. These days, automated surveillance stations, networks of wireless sensors, and satellite-based sensing systems are extensively utilized to enhance conventional sampling and laboratory analyses (Ryu 2022). This chapter explores recent advancements in water quality monitoring systems based on information and communication technology (ICT) and remote sensing.

5.2.1 Traditional Water Quality Monitoring

Monitoring of the condition of water is required for preserving the health of ecosystems and supporting the livelihoods of communities. It provides a snapshot of the condition of surface water bodies over weekly, monthly, or yearly scales (Sri Bala et al. 2025). This process involves analyzing different chemical, physical, and biological characteristics to assess whether the quality of water satisfies the requirements for its intended uses, for example, consumption, recreation, or ecosystem conservation. To protect and improve water quality, regulatory agencies at the national and state levels have established water quality criteria that define permissible pollutant limits. These criteria serve an essential part in pollution control and ensuring water security for the growing global population (Giri 2021). Traditionally, water quality monitoring relies on procedures used in laboratories. Samples of water were manually gathered from rivers, lakes, and wells and then transported to laboratories for detailed analysis. Specialized equipment and techniques were used under controlled conditions to examine the samples. The resulting data were compared with established regulations or standards to determine whether the water met the criteria for its assigned purposes (Sri Bala et al. 2025). Various techniques have been developed to keep an eye on water pollution, including the water quality index (WQI) and multivariate statistical techniques (MSTs). These tools are extremely important to the responsible handling of water resources and are commonly employed in assessments of the quality of water. MSTs help analyze complicated water quality databases, locate the origins of pollution, and distinguish between natural and human-induced factors impacting the geographical and temporal shifts in the quality of water (Varol 2020).

Communities of plankton are commonly employed as biological indicators to track ecological alterations in aquatic settings. They are effective natural bioindicators of their immediate response to changing environmental circumstances (Hemraj et al. 2017). Similarly, benthic macroinvertebrates are valuable bioindicators that often provide more precise insights into water quality changes than physicochemical or microbiological parameters. These organisms are frequently used in water quality assessments through the application of diversity and biotic indices at the community level (Wimbaningrum et al. 2016).

5.2.2 Modern Technologies for Monitoring Water Quality and Its Management

Water is a vital resource that is essential to daily life, underscoring the need for continuous monitoring to ensure its quality and usability (Essamlali et al. 2024). Previously, the condition of water was assessed by collecting field samples and analyzing them in a laboratory. However, this method, which relies on grab sampling, a process involving the collection of small water volumes, typically once a month, poses challenges. It often fails to produce reliable and representative results owing to its limited temporal and spatial scopes. Additionally, this approach is labor-intensive and financially demanding, incurring costs for both on-site sampling and laboratory analyses (Reljić et al. 2023). These challenges make consistent and integrated sampling a significant obstacle to sustainable monitoring and control of water quality.

Advancements in space science, computing technologies, and the rise of applications such as big data analytics, remote sensing, and the Internet of Things (IoT) have revolutionized water quality monitoring and prediction. These technologies enable large-scale monitoring, improved control over water treatment procedures, and timely information dissemination, which enable efficient management of water resources (Essamlali et al. 2024; Gholizadeh et al. 2016). Figure 5.2 shows the water quality management approaches.

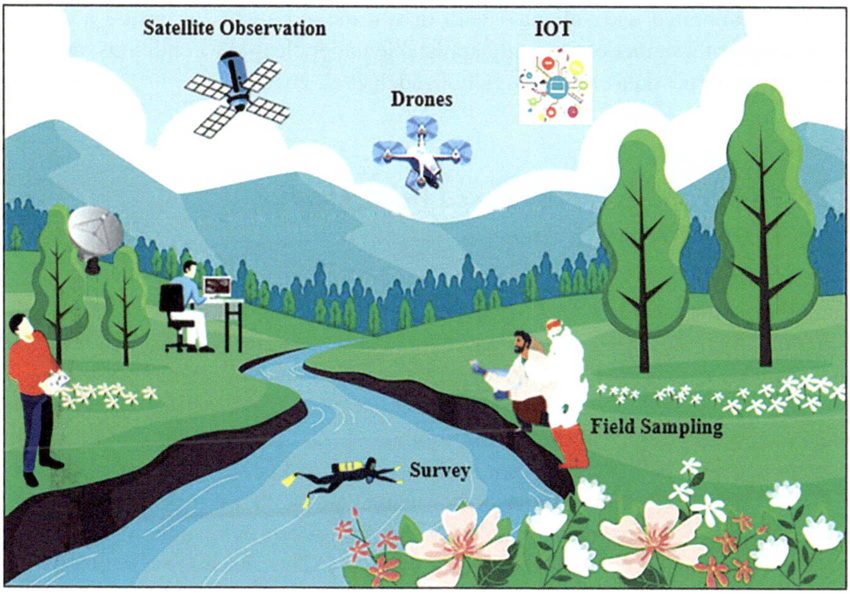

Fig. 5.2 Traditional and modern approaches for water quality management

5.2.3 Remote Sensing

Technologies for remote sensing allow for both temporal and geographical evaluations of surface water quality parameters, allowing better efficient monitoring of water bodies and quantification of water quality issues (Gholizadeh et al. 2016). This technology serves as a powerful tool for various applications, including water quality assessment, flood mapping, aquifer management, water demand modeling, water supply and level monitoring, and synoptic evaluation of soil water content. Different satellite sensors are utilized to monitor water bodies, according to the features of the research area. Quality control of water typically involves both ground-based and remote sensing components, each offering distinct advantages and limitations. The growing capabilities of remote sensing are increasingly complementing those of traditional ground-based methods. Advanced data analysis algorithms and integrated dashboards enhance decision-making for stakeholders by delivering robust, near-real-time information and highlighting critical impacts (Nagaraju et al. 2022). Remote sensing is based on the concept that water components emit or reflect varying amounts of energy across different wavelength ranges of the electromagnetic spectrum. Water pollutants alter their reflectance according to their chemical, physical, and structural features. Sensors mounted on remote sensing satellites and aerial vehicles capture this reflected or emitted energy, which is then processed into images that reveal the severity of pollution and concentration of pollutants in bodies of water. However, several key constraints must be addressed for effective application of this technique. Models for remote sensing must be carefully calibrated and validated with in situ measurements to ensure accuracy. Additionally, these models are only applicable under clear-sky conditions, as cloud cover can obstruct data collection (Gholizadeh et al. 2016).

5.2.4 Internet of Things-Based Water Quality Management System

Quality of water assessment carried out using the IoT, often referred to as smart water quality monitoring, makes it possible to observe water conditions in real time anywhere in the world. This is achieved through the integration of digital computers, services offered by the internet, connection protocols (such as TCP/IP), and mobilized sensors. Such systems have routinely been used to monitor water conditions in households, agriculture, aquaculture, municipal waste recycling, and in rivers, lakes, and other natural water bodies (Jan et al. 2021). IoT-based systems may gather information on various quality water parameters with minimal variation, allowing for qualitative comparisons and analyses. They have proven effective in tracing the impacts of contaminated water and serve as powerful tools for storing, collecting, retrieving, and managing large volumes of spatial and nonspatial data (Pant et al. 2019). This facilitates sensitivity and vulnerability analyses, thereby addressing the limitations of traditional surveillance systems. Systems that measure the condition of water based on the Internet of Things have aesthetically resolved numerous issues associated with previous methods (Jan et al. 2021). Chafa et al. (2022) designed a system for water treatment facilities that use the Internet of Things to monitor and control water quality in real time. The framework uses a microcontroller integrated with sensors to detect deviations from the water quality guidelines set forth by the World Health Organization. Danh et al. (2020) designed and implemented an Internet of Things-based system for evaluating the water condition of pangasius aquaculture farms in the delta of the Mekong. Farmers can maintain a real-time focus on important physical and chemical features with this type of technology. Moreover, it introduces an innovative automatic cleaning mechanism for sensor probes, which enhances data reliability and reduces maintenance costs.

5.2.5 Monitoring and Controlling Water Quality Using Machine Learning and Artificial Intelligence

A core component of artificial intelligence (AI) is machine learning (ML), which allows systems to learn from experience and get better on their own without the need for explicit programming (Nagaraju et al. 2024). ML models represent a transformative advancement in the research related to the monitoring and control of various engineering processes. These models utilize algorithms that are able to forecast outcomes accurately without depending on relying on complex programming. The foundation of ML lies in data mining and pattern recognition, in which algorithms are developed using a subset of the data (training set) and validated for predictive performance using another subset (testing set) (Asadollah et al. 2021).

ML has significant potential for water quality monitoring (WQM), offering tools for assessing, classifying, and predicting water quality indicators (Nagaraju et al. 2024). Technological advancements have enhanced the applicability of ML to WQM. In this field, supervised, unsupervised, reinforcement, and semi-supervised learning, there are four major types of machine learning methods that are frequently employed. Supervised learning is frequently used to identify abnormalities or changes in water quality data and to classify it according to predetermined criteria (Essamlali et al. 2024).

ML addresses complex challenges using enormous nonlinear processes or expansive combinatorial spaces that are difficult or costly to solve using traditional methods. Consequently, ML has been extensively applied across diverse domains, such as robotic control, recognition of speech, machine vision, and processing of natural languages and other emerging areas of interest (Huang et al. 2021).

5.2.6 Unmanned Vehicle Technology for Monitoring Water Quality

Unmanned vehicles are designed to operate in a variety of settings, such as water, air, and land. Unmanned surface vehicles (USVs) are utilized for operation on the water's surface, airborne variants are known as unmanned aircraft (UAVs), and unmanned underwater vehicles (UUVs) are vehicles specifically designed for adoption in underwater environments (Wibisono et al. 2023). Recently, these technologies have gained significant traction in aquaculture and agriculture, especially for the management and observation of fish owing to their increased accessibility and cost-effectiveness.

Drones, flying robots, and unmanned aircraft systems are other names for unmanned aerial vehicles, or UAVs, and have emerged as globally favored methods for efficiently mapping and monitoring water resources at local scales because of their speed, accuracy, and cost-effectiveness (Sibanda et al. 2021; Acharya et al. 2021). These UAV-based techniques offer ultra-high-resolution data, often within the centimeter range, and allow for flexible temporal data acquisition, limited only by sea surface weather conditions, as cloud cover does not hinder data collection. UAVs are significantly more affordable than other remote-sensing tools for acquisition and operation. In order to determine parameters related to water quality including sea surface temperature (SST), total suspended solids (TSS), chlorophyll (Chl-a), total nitrogen, and total phosphorous, lightweight sensors installed on UAVs can collect data across optical RGB, thermal, multispectral, and LiDAR platforms (Ryu 2022). The results of these measurements are similar to those obtained through satellites. Improvements in wireless networks, robotics, sensor tools, and computational image processing systems have enhanced UAV capacities, supporting in situ observation in real time and extending the spatial resolution of water quality assessment (Acharya et al. 2021).

USVs, also referred to as autonomous surface craft or marine drones, are boats or ships designed to operate on or beneath the surface of water bodies without requiring human operators (Ubina and Cheng 2022). Initially developed for unmanned activities, like data collection and monitoring the environment, USVs have gained widespread use in military applications and are increasingly adopted within the marine and aquatic science sectors. Their appeal lies in their ability to minimize safety risks, lower labor expenses, and improve measurement performance (Erena et al. 2019). To be effective, USVs must be durable and easy to operate in diverse environments. They are capable of approaching targets to capture high-resolution images and perform rapid real-time missions. Their mobility and ability to cover extensive areas with precise sensors make them advantageous instruments for various applications, observing the quality of the water and conducting bathymetric inspections in the environmental sector (Castellini et al. 2020). Additionally, USVs can conduct prolonged monitoring tasks because they can be equipped with heavier batteries that offer higher energy capacities.

Underwater unmanned vehicles (UUVs) are robotic devices that operate beneath the water's surface, also known as underwater drones or sea gliders. Vehicles like these are separated into two main groups: autonomous underwater vehicles (AUVs) and remotely operated vehicles (ROVs) (Kwon et al. 2023). Both ROVs and AUVs are designed to be waterproof and submersible, equipped with cameras for capturing images and videos, as well as various sensors for collecting information about the quality of the water. The sensors in ROVs can gather information on factors including the depth of the water, temperature, and physical, chemical, and biological characteristics (Ubina and Cheng 2022). With advancements in UUV and sensor technologies, these vehicles have become widely used for water quality assessment, as they can be equipped with diverse water samplers and sensors. UUVs are capable of reaching extreme depths, making them suitable for both water quality measurements and underwater tasks (Kwon et al. 2023).

5.2.7 *Citizen Science-Based Water Quality Monitoring*

Citizen science is an unconventional method that involves public participation in various scientific endeavors, including environmental monitoring and reporting. This method fosters collaboration between experts and amateurs, potentially leading to joint research design (Wu et al. 2022). Such initiatives align the interests of local and academic communities toward shared objectives, such as ensuring the sustainability of nearby resources. The global acceptance of citizen science has significantly increased in the last few centuries. This approach offers substantial potential, as it can address knowledge gaps in specific areas, enhance environmental consciousness to stimulate behavioral changes, supplement traditional monitoring with cost-effective alternatives, and attract corporate funding and backing (Yevenes et al. 2022). The inclusion of the general public in monitoring water quantity and quality is considered advantageous for several reasons. A primary benefit is the ability to gather data across broader spatial and temporal scales compared to conventional scientific methods. Additionally, citizen science is viewed as a means to achieve the monitoring standards necessary for attaining water-related sustainable development goals. Furthermore, it can improve scientific knowledge, heighten awareness of the environment, and strengthen the community's ability to safeguard the environment (Ramírez et al. 2023). Table 5.1 shows the application of various approaches and their significance.

5.3 Limitations of Water Quality Monitoring Approaches

Conventional techniques for assessing the quality of water, such as laboratory testing and observation in the field, are frequently used. The precision and reliability of traditional water quality monitoring techniques, such as laboratory analysis and field testing, are widely acknowledged; however, these methods are often limited by factors such as low spatiotemporal resolution, high equipment costs, and time-consuming, labor-intensive procedures. These restrictions make it challenging to do comprehensive water quality monitoring and gather data instantaneously.

New methods, such as machine learning algorithms and remote sensing, have therefore gained popularity because of their ability to get around these restrictions and provide better and more effective water quality monitoring solutions. These techniques, which involve collecting and analyzing data from large bodies of water, provide a more comprehensive understanding of water quality with higher spatio-temporal resolution, lower costs, and reduced labor requirements. Conventional testing and sampling methods to monitor water quality have opened the way to remote sensing, Internet of Things (IoT)-based sensors, unmanned vehicle technology, machine learning, and artificial intelligence (Nagaraju et al. 2023).

Satellite remote sensing contributes immensely to the scientific monitoring of the quality of water because it enables global monitoring of water quality indicators, and researchers along with users can obtain broad and constant spatial coverage of bodies of water. Using satellite technology and algorithms for learning to predict the quality of water presents substantial obstacles, particularly in terms of data quality, creating models, and execution. Quality in data is a critical concern since higher-resolution images without shadows are sometimes affected by conditions in the atmosphere like haze and dense clouds, resulting in error. Satellite data's temporal resolution may not correspond with monitoring demands, and inconsistent geographical resolution across platforms complicates data integration (Deng et al. 2024). This is largely due to a shortage of expert skills in geospatial analysis of data, restricted public engagement, and a shortage of training data and code resources

Table 5.1 Review of the application of different approaches in water quality management and monitoring

Research	Purpose	Satellite sensor and techniques	References
Coastal surface water quality	Retrieve dissolved oxygen concentration	MODIS & VIIRS	Kim et al. (2020)
Assessment of changes in lake water quality characteristics	Total phosphorus, total suspended solids, chlorophyll-a, and Kjeldahl nitrogen content	Landsat (TM), Landsat (OLI)	Hajigholizadeh et al. (2021)
Coastal lagoon water quality	Turbidity and chlorophyll-a	Landsat and sentinel	Caballero et al. (2022)
Water quality retrieval	Water quality indicators include chlorophyll-a, total suspended concentration, aggregate total nitrogen, total phosphorus, chemical oxygen demand, and colored dissolved organic matter	Review	Yang et al. (2022)
Water quality	Dissolved oxygen	Algorithms (random forest and support vector machine) applied to sentinel imagery	Salas et al. (2022)
Monitoring aquaculture water quality	Physicochemical variables of the pond water	IoT	Danh et al. (2020)
Examine the quality of the water	Water quality parameters	IoT system with wireless multi-sensors	Ighalo et al. (2021) and Tsai et al. (2021)
Monitoring and regulating fishpond water quality	An intelligent fish farming system based on the Internet of Things uses predictions to control water quality automatically	A module for tracking aquatic products and intelligent management	Gao et al. (2019)
System for measuring surface quality of water	Water quality parameters	IoT and intelligent technology	Zaidi Farouk et al. (2023)
River water quality	Dissolved oxygen	Random forest (RF), gradient boosting algorithms (GB), multilayer perceptrons (MLP), support vector regression (SVR), and SHapley additive explanations (SHAP)	Li et al. (2024)
Water quality	Predicting water and groundwater quality indices	Decision trees (DT), adaptive neuro-fuzzy inference system, and kernel Gaussian learning	Jibrin et al. (2024)

(continued)

Table 5.1 (continued)

Research	Purpose	Satellite sensor and techniques	References
Assessment of river quality	Physicochemical variables and heavy metal	ANN, SVM, RF, DT, and DL	Gottumukkala et al. (2024)
Overall quality of water	Physicochemical variables	ANN and LSTM	Rana et al. (2023)
Retrieve water quality	Chlorophyll-a and total suspended solids	Multispectral sensor with five bands installed on a UAV	Román et al. (2022)
Water quality monitoring bodies of inland water	Physicochemical parameters	Data from UAVs and satellites for surveillance purposes	Wasehun et al. (2024)
Real-time monitoring	Hazardous algal blooms and the quality of water	UAV system	Hagh et al. (2024)

that are openly accessible. These flaws impede development in determining the condition of water through remote sensing while also perpetuating the false belief that remote sensing estimations are unreliable (Nagaraju et al. 2023).

Monitoring the quality of surface water via IoT or smart systems is more efficient and cost-effective than traditional approaches, allowing for continuous online monitoring. However, the real-time water quality monitoring system has limitations, including limited capacity for transmission, retention of information, battery life, network assets, and durability. When implementing the system, it's important to consider cyber-attacks as a potential hazard. The absence of market-available sensors for analyzing physical, chemical, and biological factors poses a significant hurdle for IoT-based inspections of water quality (Zaidi Farouk et al. 2023). Although unmanned vehicle systems improve the precision and usability of remote monitoring of water quality, they have limits in terms of sensor technology and data processing algorithms. Autonomous surface vehicles and drones are examples of unmanned vehicles that are frequently limited by their payload capacity and battery life. The adoption of drones may be limited by regulatory limits on their use in specific locations and difficulties in operating in inclement weather. Widespread acceptance is further hampered by the hefty startup expenses and technological know-how needed for operation.

Community-based monitoring (CBM), a participatory method of monitoring water quality, has gained popularity despite facing several challenges. CBM typically encounters three main types of issues: data collection, data utilization, and organizational difficulties. At the organizational level, challenges include a lack of volunteer motivation, limited networking opportunities, inadequate funding, and restricted access to information. Data collection problems may arise due to limited participant objectivity, unreliable data, and fragmented datasets, as identified by Whitelaw et al. (2003). Moreover, it has been suggested that decision-makers often overlook data gathered by community organizations due to concerns about its completeness, comparability, and credibility (Conrad and Hilchey 2010).

5.4 Recommendations to Enhance Water Quality Monitoring Approaches

It is imperative to combine modern technologies and innovative ideas with conventional approaches in order to overcome these limitations. Regarding water quality based on remote monitoring, typical field sampling dates should align with either satellite crossovers or remote sensing surveys. This method increases the precision and trustworthiness of the quality of water observation initiatives by optimizing the use of remote sensing data and sampling at the right moment (Nagaraju et al. 2023). Providing these datasets and code sources accessible to the general public encourages greater involvement of the public in water science that relies on remotely sensed data. It additionally serves an important role in promoting the creation of models and assurance, boosting precision, and enhancing public opinion. Using freely available tools, academics and professionals are able to speed development, improve reproducibility, and collaborate to successfully address difficult water quality concerns. These initiatives not only enable independent scholars but also foster cooperation, support training programs for users, raise awareness in society, and make significant contributions to the worldwide advancement of algorithms. They promote affordable research and transparency, enabling both researchers and professionals to make substantial progress in satellite-based water condition monitoring. IoT-based devices can be enhanced by implementing energy-efficient parameter sensors, enhancing cybersecurity protocols, and improving communication stability, measurement consistency, and reliability (Zaidi Farouk et al. 2023). Finally, to enhance community-based monitoring and inspire involvement, an educational program centered on environmental monitoring should be created. The benefits for participants, such as individual fulfillment and the development of data-driven policies that benefit all stakeholders, should be highlighted in this program, which is being led by nongovernmental organizations, governmental entities, and environmental organizations. The issue of unused data produced by community scientific initiatives can be addressed by connecting regional community science platforms to international networks and databases, such as iNaturalist and the Canadian National Conservation Initiative (Conrad and Hilchey 2010).

5.5 Conclusions

In conclusion, society, the government, and nonpublic sectors should all take seriously the detrimental issue of deteriorating water quality. Effective river delta management necessitates a comprehensive evaluation of past and current changes, including land usage and cover changes, ecosystem changes, and alterations in river and coastal geomorphology. This multifaceted assessment is challenging to accomplish solely through conventional land-based survey methods. Earth observation techniques can significantly contribute by providing precise data on these processes.

Recent technological advancements in information and communication technology and remote sensing have enhanced surveillance of water quality. These improvements have substantially increased our capacity to describe dynamic hydrologic characteristics at suitable geographical and temporal scales. Consequently, both statistical and mechanistic models for monitoring water quality trends have been refined at local, watershed, and regional scales across freshwater, estuarine, and marine systems. Furthermore, these advancements have significantly increased the ability to identify hydrologic changes, which is crucial for mechanisms for timely warning and swift responses to damaging algal blooms. These technological developments offer methods and data for detailed, affordable, continuous, and standardized monitoring of water diversity and quality properties from local to continental scales, supporting decision-making processes.

Thus, the review aims to give an overview of traditional and modern methods while highlighting a comparative analysis between them. Water quality monitoring methods such as remote sensing, IoT approach, machine learning and artificial intelligence, and unmanned vehicle technology are studied. The review emphasizes how modern innovations can support water quality monitoring and promote knowledge-based decision-making for sustainable water resource management.

References

Acharya BS, Bhandari M, Bandini F, Pizarro A, Perks M, Joshi DR et al (2021) Unmanned aerial vehicles in hydrology and water management: applications, challenges, and perspectives. Water Resour Res 57(11):e2021WR029925

Anthony EJ (2015) Wave influence in the construction, shaping and destruction of river deltas: a review. Mar Geol 361:53–78

Asadollah SBHS, Sharafati A, Motta D, Yaseen ZM (2021) River water quality index prediction and uncertainty analysis: a comparative study of machine learning models. J Environ Chem Eng 9(1):104599

Bhadra T, Hazra S, Ray SS, Barman BC (2020) Assessing the groundwater quality of the coastal aquifers of a vulnerable delta: a case study of the Sundarban Biosphere Reserve, India. Groundw Sustain Dev 11:100438

Caballero I, Roca M, Santos-Echeandía J, Bernárdez P, Navarro G (2022) Use of the sentinel-2 and landsat-8 satellites for water quality monitoring: an early warning tool in the mar menor coastal lagoon. Remote Sens 14(12):2744

Castellini A, Bloisi D, Blum J, Masillo F, Farinelli A (2020) Multivariate sensor signals collected by aquatic drones involved in water monitoring: a complete dataset. Data Brief 30:105436

Chafa AT, Chirinda GP, Matope S (2022) Design of a real–time water quality monitoring and control system using internet of things (IoT). Cogent Eng 9(1):2143054

Conrad CC, Hilchey KG (2010) A review of citizen science and community-based environmental monitoring: issues and opportunities. Environ Monit Assess 176(1–4):273–291. https://doi.org/10.1007/s10661-010-1582-5

Danh LVQ, Dung DVM, Danh TH, Ngon NC (2020) Design and deployment of an IoT-based water quality monitoring system for aquaculture in Mekong Delta. Int J Mech Eng Robot Res 9(8):1170–1175

Deng Y, Zhang Y, Pan D, Yang SX, Gharabaghi B (2024) Review of recent advances in remote sensing and machine learning methods for Lake water quality management. Remote Sens 16(22):4196–4196. https://doi.org/10.3390/rs16224196

Erena M, Atenza JF, García-Galiano S, Domínguez JA, Bernabé JM (2019) Use of drones for the topo-bathymetric monitoring of the reservoirs of the Segura River Basin. Water 11(3):445

Essamlali I, Nhaila H, El Khaili M (2024) Advances in machine learning and IoT for water quality monitoring: a comprehensive review. Heliyon 10:e27920

Feyen L, Ciscar Martinez JC, Gosling S, Ibarreta Ruiz D, Soria Ramirez A, Dosio A et al (2020) Climate change impacts and adaptation in Europe. JRC PESETA IV final report (No. JRC119178). Joint Research Centre (Seville site)

Gao G, Xiao K, Chen M (2019) An intelligent IoT-based control and traceability system to forecast and maintain water quality in freshwater fish farms. Comput Electron Agric 166:105013

Gholizadeh MH, Melesse AM, Reddi L (2016) A comprehensive review on water quality parameters estimation using remote sensing techniques. Sensors 16(8):1298

Giri S (2021) Water quality prospective in twenty first century: status of water quality in major river basins, contemporary strategies and impediments: a review. Environ Pollut 271:116332

Gottumukkala SB, Thotakura VN, Gvr SR, Chinta DP, Park R (2024) Balancing aquaculture and estuarine ecosystems: machine learning–based water quality indices for effective management. Environ Sci Pollut Res 31:1–17

Hagh SF, Amngostar P, Zylka A, Zimmerman M, Cresanti L, Karins S, O'Neil-Dunne JP, Ritz K, Williams CJ, Morales-Williams AM, Huston D, Xia T (2024) Autonomous UAV-mounted LoRaWAN system for real-time monitoring of harmful algal blooms (HABs) and water quality. IEEE Sensors J 24(7):11414–11424. https://doi.org/10.1109/jsen.2024.3364142

Hajigholizadeh M, Moncada A, Kent S, Melesse AM (2021) Land–lake linkage and remote sensing application in water quality monitoring in Lake Okeechobee, Florida, USA. Land 10(2):147

Hemraj DA, Hossain MA, Ye Q, Qin JG, Leterme SC (2017) Plankton bioindicators of environmental conditions in coastal lagoons. Estuar Coast Shelf Sci 184:102–114

Huang R, Ma C, Ma J, Huangfu X, He Q (2021) Machine learning in natural and engineered water systems. Water Res 205:117666

Ighalo JO, Adeniyi AG, Marques G (2021) Internet of things for water quality monitoring and assessment: a comprehensive review. In: Artificial intelligence for sustainable development: theory, practice and future applications, pp 245–259

Jan F, Min-Allah N, Düştegör D (2021) Iot based smart water quality monitoring: recent techniques, trends and challenges for domestic applications. Water 13(13):1729

Jibrin AM, Al-Suwaiyan M, Aldrees A, Dan'azumi S, Usman J, Abba SI et al (2024) Machine learning predictive insight of water pollution and groundwater quality in the Eastern Province of Saudi Arabia. Sci Rep 14(1):20031

Kim YH, Son S, Kim HC, Kim B, Park YG, Nam J, Ryu J (2020) Application of satellite remote sensing in monitoring dissolved oxygen variabilities: a case study for coastal waters in Korea. Environ Int 134:105301

Kontopoulos C, Kitsiou D, Chlorokostas S, Charalampopoulou B, Grammalidis N, Tzepkenlis A (2022) Identification and analysis of long-term changes in river deltas and riparian zones using time-series multispectral data. Euro-Mediterr J Environ Integr 7(4):553–564

Kuenzer C, Heimhuber V, Huth J, Dech S (2019) Remote sensing for the quantification of land surface dynamics in large river delta regions—a review. Remote Sens 11(17):1985

Kwon DY, Kim J, Park S, Hong S (2023) Advancements of remote data acquisition and processing in unmanned vehicle technologies for water quality monitoring: an extensive review. Chemosphere 343:140198–140198. https://doi.org/10.1016/j.chemosphere.2023.140198

Li S, Qasem SN, Band SS, Ameri R, Pai HT, Mehdizadeh S (2024) Explainable machine learning models for estimating daily dissolved oxygen concentration of the Tualatin River. Eng Appl Comput Fluid Mech 18(1):2304094

Liu Y, Li J, Sun C, Wang X, Tian P, Chen L et al (2023) Thirty-year changes of the coastlines, wetlands, and ecosystem services in the Asia major deltas. J Environ Manag 326:116675

Loucks DP (2019) Developed river deltas: are they sustainable? Environ Res Lett 14(11):113004

Nagaraju TV, Malegole SB, Chaudhary B, Ravindran G (2022) Assessment of environmental impact of aquaculture ponds in the western delta region of Andhra Pradesh. Sustain For 14(20):13035

Nagaraju TV, Sunil BM, Chaudhary B, Prasad CD, Gobinath R (2023) Prediction of ammonia contaminants in the aquaculture ponds using soft computing coupled with wavelet analysis. Environ Pollut 331:121924

Nagaraju TV, Malegole SB, Chaudhary B, Ravindran G, Chitturi P, Chinta DP (2024) Novel assessment tools for inland aquaculture in the western Godavari delta region of Andhra Pradesh. Environ Sci Pollut Res 31(25):36275–36290

Pant D, Bhatt A, Khan M, Nautiyal OP, Adhikari P (2019, November) Automated IoT based smart water quality assessment system. In: 2019 8th international conference system modeling and advancement in research trends (SMART). IEEE, pp 98–104

Peng Y, Welden N, Renaud FG (2023) A framework for integrating ecosystem services indicators into vulnerability and risk assessments of deltaic social-ecological systems. J Environ Manag 326:116682

Ramírez SB, van Meerveld I, Seibert J (2023) Citizen science approaches for water quality measurements. Sci Total Environ 165436:165436

Rana R, Kalia A, Boora A, Alfaisal FM, Alharbi RS, Berwal P et al (2023) Artificial intelligence for surface water quality evaluation, monitoring and assessment. Water 15(22):3919

Reljić M, Romić M, Romić D, Gilja G, Mornar V, Ondrasek G et al (2023) Advanced continuous monitoring system—tools for water resource management and decision support system in salt affected delta. Agriculture 13(2):369

Renaud FG, Syvitski JP, Sebesvari Z, Werners SE, Kremer H, Kuenzer C et al (2013) Tipping from the Holocene to the Anthropocene: how threatened are major world deltas? Curr Opin Environ Sustain 5(6):644–654

Renaud FG, Szabo S, Matthews Z (2016) Sustainable deltas: livelihoods, ecosystem services, and policy implications. Sustain Sci 11:519–523

Román A, Tovar-Sánchez A, Gauci A, Deidun A, Caballero I, Colica E et al (2022) Water-quality monitoring with a UAV-mounted multispectral camera in coastal waters. Remote Sens 15(1):237

Ryu JH (2022) UAS-based real-time water quality monitoring, sampling, and visualization platform (UASWQP). HardwareX 11:e00277

Salas EAL, Kumaran SS, Partee EB, Willis LP, Mitchell K (2022) Potential of mapping dissolved oxygen in the little Miami River using Sentinel-2 images and machine learning algorithms. Remote Sens Appl Soc Environ 26:100759

Scown MW, Dunn FE, Dekker SC, van Vuuren DP, Karabil S, Sutanudjaja EH et al (2023) Global change scenarios in coastal river deltas and their sustainable development implications. Glob Environ Chang 82:102736

Sibanda M, Mutanga O, Chimonyo VG, Clulow AD, Shoko C, Mazvimavi D et al (2021) Application of drone technologies in surface water resources monitoring and assessment: a systematic review of progress, challenges, and opportunities in the global south. Drones 5(3):84

Sri Bala G, Nagaraju TV, Krishnam Raju GLV, Rambabu T, Harish Kumar Varma G (2025) Advancing Estuarine Aquaculture with Remote Sensing for Water Management. In Inland Aquaculture Sustainability and Effective Water Management Strategies (pp 87–97). Springer, Cham

Tsai HL, Lin JY, Lyu WH (2021, November) Design and evaluation of wireless multi-sensor IoT system for monitoring water quality of freshwater aquaculture. In: 2021 international automatic control conference (CACS). IEEE, pp 1–6

Ubina NA, Cheng SC (2022) A review of unmanned system technologies with its application to aquaculture farm monitoring and management. Drones 6(1):12

Varol M (2020) Use of water quality index and multivariate statistical methods for the evaluation of water quality of a stream affected by multiple stressors: a case study. Environ Pollut 266:115417

Wasehun ET, Hashemi Beni L, Di Vittorio CA (2024) UAV and satellite remote sensing for inland water quality assessments: a literature review. Environ Monit Assess 196(3):277

Wibisono A, Piran MJ, Song HK, Lee BM (2023) A survey on unmanned underwater vehicles: challenges, enabling technologies, and future research directions. Sensors 23(17):7321

Wimbaningrum R, Indriyani S, Retnaningdyah C, Arisoesilaningsih E (2016) Monitoring water quality using biotic indices of benthic macroinvertebrates along surfaces water ecosystems in some tourism areas in East Java, Indonesia. J Indones Tour Dev Stud 4(2):81

Wu Y, Washbourne C, Haklay M (2022) Citizen science in China's water resources monitoring: current status and future prospects. Int J Sustain Dev World Ecol 29(3):277–290

Wu S, Zhou X, Reyns J, Yamazaki D, Yin J, Li X (2024) Climate change and urban sprawl: unveiling the escalating flood risks in river deltas with a deep dive into the GBM river delta. Sci Total Environ 947:174703

Yang H, Kong J, Hu H, Du Y, Gao M, Chen F (2022) A review of remote sensing for water quality retrieval: progress and challenges. Remote Sens 14(8):1770

Yevenes MA, Pereira H, Bermudez R (2022) Citizen science as a co-creative measure to water quality: chemical data and local participation in a rural territory. Front Environ Sci 10:940778

Zaidi Farouk MIH, Jamil Z, Abdul Latip MF (2023) Towards online surface water quality monitoring technology: a review. Environ Res 238:117147. https://doi.org/10.1016/j.envres.2023.117147

Chapter 6
Occurrence and Presence of Ubiquitous Microplastics and Their Effects on Delta Ecosystems Around the World

Satish Kumar Kolluru and Vinoth Kumar Raja

Abstract The ubiquitous microplastics (MPs), having sizes smaller than 5 mm, have become abundant in aquatic environments, raising concerns about their environmental and health implications. This chapter begins by detailing the occurrence of MPs in delta regions, highlighting their pervasive presence in water bodies, sediments, and biota. MPs pose a significant challenge to delta ecosystems around the world, affecting biodiversity, human health, and food security via a variety of pathways that include urban runoff, industrial discharges, agricultural practices, degradation of larger plastics, atmospheric deposition, synthetic textiles, tire wear, and other sources. It discusses their impact on aquatic organisms, including fish and invertebrates, which can ingest these particles, leading to physical harm and toxicological effects. Protecting delta ecosystems from the contamination of MPs is critical for maintaining ecological integrity and ensuring the sustenance of numerous individuals who depend on these areas for nourishment and economic activities. In conclusion, this chapter underscores the urgent prerequisite for a comprehensive study to analyze the dynamics of MP contamination in delta ecosystems. It advocates for effective management strategies to mitigate this growing threat, emphasizing the importance of policy interventions, public awareness campaigns, and sustainable practices to protect these crucial environments for future generations.

Keywords Microplastics · Delta ecosystem · Biodiversity · Microplastics ingestion · Treatment · Control strategies

S. K. Kolluru · V. K. Raja (✉)
Membrane Research Laboratory, Department of Chemical Engineering, National Institute of Technology Andhra Pradesh, Tadepalligudem, Andhra Pradesh, India
e-mail: vinoth@nitandhra.ac.in

T. V. Nagaraju, G. S. Bala (eds.), *Delta Ecosystems*, Environmental Contamination Remediation and Management, https://doi.org/10.1007/978-3-031-97920-0_6

6.1 Introduction

Delta ecosystems are among the most productive and diverse environments in the world, providing crucial resources for millions of people. Deltas contribute significantly to global food security by providing fertile agricultural land and abundant fisheries. Deltas are landforms created at the entry of a water body that connects a part of water to a large water body. Deltas support diverse flora and fauna, resulting in rich biodiversity. They also host a variety of species, including fish, birds, and plants, making them critical for maintaining ecological balance. Estuaries often support large human populations due to their agricultural potential and access to freshwater resources (Scown et al. 2023).

The deltas are distributed across the globe, such as the Ganges–Brahmaputra Delta, which spans most of Bangladesh and India's West Bengal, which is the biggest delta (Allison 1998; Islam and Gnauck 2008). The Sundarbans, a UNESCO World Heritage Site located at the mouth of the Ganges River, contains the world's largest mangrove forest (Neelavannan et al. 2023). Other significant river deltas include the Sacramento–San Joaquin and Mississippi River deltas in the USA, the Nile Delta in Egypt, the Yangtze Delta in China, the Selenga River Delta in Russia (the largest delta discharging into a freshwater body, the Baikal lake), the Tiber Delta in Italy, the Ebro Delta in Spain, the Yukon Delta in eastern Alaska, the Amazon Delta in Brazil, the Rhine–Meuse–Scheldt Delta in the Netherlands, the Irrawaddy Delta in Myanmar, and the Seine River Delta in France (Syvitski et al. 2022). The satellite images of different deltas are illustrated in Fig. 6.1.

6.2 Threats to Deltas

Deltas are vital ecosystems that face a myriad of threats, which can significantly impact their health, functionality, and the communities that rely on them. Crumbling of delta habitats may cause huge hindrances to comprehensive sustainable growth. The key threats to delta ecosystems are climate change, land subsidence, population growth and urbanization, pollution, sediment loss, salinization, ecosystem degradation, economic constraints, and the entry of MPs into the delta ecosystem (Syvitski

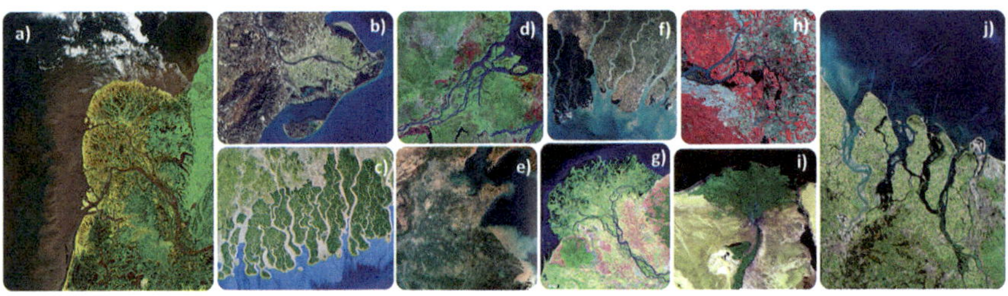

Fig. 6.1 The satellite images of deltas around the globe: (**a**) Yukon, (**b**) Ebro, (**c**) Sundarbans, (**d**) Amazon, (**e**) Shanghai, (**f**) Irrawaddy, (**g**) Selenga, (**h**) Sacramento, (**i**) Nile, and (**j**) Rhine–Meuse–Scheldt. (Source: ©ESA and NASA Worldview Snapshots)

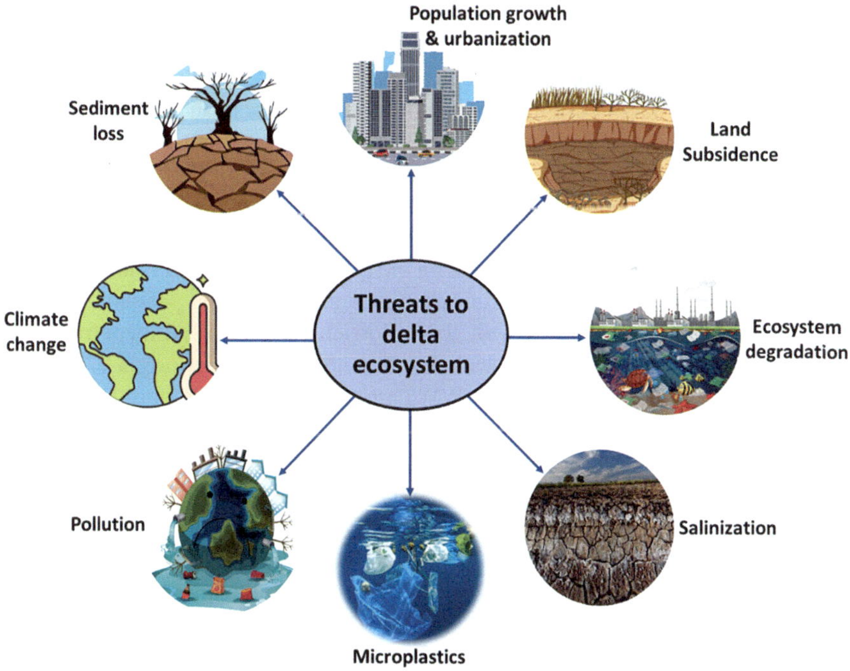

Fig. 6.2 Threats that are causing destruction to the delta ecosystem

2008). The degradation of mangrove forests and wetlands due to urban development, pollution, and climate change reduces biodiversity and the ecosystem services these habitats provide. Mangroves are crucial for coastal protection and carbon sequestration, and they serve as nursery grounds for many fish species (Islam and Gnauck 2008; Sarker et al. 2022). The major threat to the delta ecosystem has been pollution, which is due to agricultural runoff, industrial discharges, and urban waste contributing to water pollution in delta regions. Contaminants such as pesticides, heavy metals, and other pollutants devalue water standards and harm aquatic creatures, impacting fisheries and overall ecosystem well-being (Schneider and Asch 2020; Kuenzer and Renaud 2012). In the past decade, a new threat has been destroying the quality of the delta ecosystem at the micro level called MPs, which have become a macro peril to the delta ecosystem as it has entered the ecosystem food chain (Gao et al. 2022; Kumar et al. 2022). The threats that are causing threat and destruction to delta ecosystems are presented in Fig. 6.2.

6.3 MPs in Delta Ecosystems

Delta ecosystems are among the most productive and diverse environments in the world, providing crucial resources for millions of people. However, these vital regions are increasingly threatened by pollution, particularly from MPs. This chapter explores the proximity of MPs in delta ecosystems globally, examining their sources, impacts on the milieu and human vigor, and potential treatment methods to mitigate their effects (Klein et al. 2018; Ramesh et al. 2024; Vivekanand et al. 2021).

MPs have a number of origins, ranging from huge plastic trash that further breaks down to tiny fragments. These minute fragments, which are 5 mm or smaller in size, are called MPs. These have caused great concern to the environment worldwide due to their extensive and prospective ecological effects (Oßmann et al. 2018; Chen et al. 2021; Kankanige and Babel 2021). They are produced by many origins and either primary or secondary MPs based on the type of production. Primary MPs are generated on purpose and utilized in factories, e.g., textiles and personal care and cosmetic products (PCCP) (Lin et al. 2023; Borriello et al. 2012). Fibers from synthetic clothing like nylon and polyester are the other substantial origin of primary MPs, and by washing machines and WWT units, they penetrate into the environment straightforwardly (Kim et al. 2024). However, due to other environmental factors, huge plastic particles disintegrate to produce secondary MPs. Every kind of MP presents critical environmental anxiety because of their perseverance, capability, and mobility to build up in the ecosystem (Browne et al. 2011). MPs enter into the food hierarchy and become a concern for health risks for humans as well as animals (Sigler 2014). Also, microbeads, which are very minute spherical plastics, are used in the making of beauty products such as detergents and kinds of toothpaste (Kalčíková et al. 2017). These minute fragments smoothly pass through water purification facilities and end up in water bodies, causing impeccable harm to water animals (Cera et al. 2023).

Microbeads are not really a problem today; they have existed for 50 years, when natural products were replaced with plastics in cosmetics and personal care products, as stated by UNEP (United Nations Environment Programme). MPs are of different origins and consist of huge plastic fragments, which are fragmented into resin pellets utilized in plastic production or in the shape of microbeads, which are tiny, produced plastic beads utilized in healthcare and cosmetics (Perren et al. 2018; Hoang et al. 2022).

6.4 Status of MPs in Global Delta Ecosystems

6.4.1 Sundarbans

MPs have been detected in sediments, surface water, and aquatic organisms within the Sundarbans Delta, indicating a significant pollution load. Over 40 lakh tons of plastics are discharged yearly and reach the Bay of Bengal from several water channels in Bangladesh and India (Parvin et al. 2021). MPs pose substantial ecological risks, including trophic transfer and bioaccumulation across different species. Increased concentrations of MPs were observed with higher trophic levels, indicating potential harm to the food web. One study indicated that MPs can lead to physical damage to biota and chemical toxicity and also could change the phytoplankton dynamics (Kumar et al. 2022; Kumar et al. 2021).

Shetu et al. (2023) probed the presence of plastic fragments in frogs from the Bengal Delta, an ecologically rich area significantly impacted by plastic pollution, and found MPs in 90% of sampled frogs, with *Asian Common Toads* showing the highest concentration (Shetu et al. 2023). This high prevalence highlights serious pollution levels in the delta. MPs identified included fibers, fragments, beads, and foams, with fibers being the most common, originating from consumer products, textiles, and local industrial sources (Dris et al. 2017). The study stressed the ecological risks, suggesting that frogs may serve as bioindicators of MP pollution due to their role in the food chain, potentially transferring MPs to higher trophic levels (He et al. 2022). Neelavannan et al. (2023) studied the Ganges–Brahmaputra Delta, focusing on the Hooghly River and the Sundarbans Biosphere Reserve for MPs. They found high concentrations of MPs averaging 718 fragments/m^3 and 428 fragments/kg in water and sediment samples, respectively. Predominant polymers identified were high-density polyethylene (HDPE), polyoxymethylene (POM), and polyphenylene sulfide (PPS) (Neelavannan and Sen 2023).

Sarker et al. (2022) evaluated the trophic shift and MP biomagnification across different species within the Sundarbans mangrove ecosystem. MPs were found across all studied species (crab, shrimp, mollusks, turtle, fish, and shark), showing the highest contamination, indicating the magnitude of MP contamination within the bionetwork. This analysis confirmed that MPs bio-magnify across trophic levels, with an 8.16-fold increase from primary to quaternary consumers, showing that MPs concentrate as they move up the food chain. The study spotted that predatory, filter-feeding, and browsing species had a higher ingestion rate of MPs, suggesting that feeding strategies influence MP accumulation (Sarker et al. 2022). Nawar et al. (2023) provided an in-depth characterization of MP contamination in the Pasur River of the Sundarbans, Bangladesh, converging on water, fish, and soil samples (Nawar et al. 2023). MPs were present across analyzed trials, exposing an average of 2.66×10^3 fragments/L in water and 1.57×10^5 fragments/kg in sediments. MPs were found in the GI (gastrointestinal) tract and muscle tissue of various fish species, with benthic species showing the highest MP accumulation (Nawar et al. 2023). This suggests significant MP ingestion through trophic levels, posing risks to both ecosystem and human health. The possible primary and secondary sources of MPs in the Sundarbans are illustrated in Fig. 6.3.

6.4.2 *Ebro Delta*

Simon-Sánchez et al. (2019) examined MP accumulation in the Ebro River Delta, positioning river deltas as significant hotspots for MP pollution (Llorca et al. 2021). MPs were found across different matrices, with fibers as the predominant type. Estuarine sediments showed particularly high MP accumulation, suggesting that deltas may act as both sources and sinks for MPs. The delta's salt-wedge estuary dynamics likely facilitate MP deposition in sediments, where lower flow velocity in the salt-wedge zone contributes to the settling of MPs. The study underscored the need for more comprehensive monitoring and emphasized the importance of delta

Fig. 6.3 An illustration showing different sources of MPs in the Sundarbans

regions in the global distribution and fate of MPs. Llorca et al. (2021) used a pioneering double suspect-screening method for identifying polymers, additives, and adsorbed contaminants in micro and nanoplastics (NPL/MPs) within environmental waters from the Ebro Delta (Simon-Sánchez et al. 2019). With polymer concentrations reaching 7000 ng/L in some areas, this work stressed the role of NPL/MPs as vectors for pollutants, demonstrating their potential ecological impact in both freshwater and seawater environments, dominated by polymers like PP, PE, and PS (Llorca et al. 2021; Simon-Sánchez et al. 2019).

6.4.3 Yellow River Delta

Duan et al. (2020) found out that MPs were detected across all study areas, including the protected zones with minimal human activity, indicating significant contamination throughout the delta, and also suggested that direct human activities substantially increased MP contamination. Polyethylene terephthalate (PET) was consistently detected and correlated with MP levels, with concentrations ranging from 536 to 660 μg/kg. Polycarbonate (PC) concentrations ranged from 83.9 to 196 μg/kg, suggesting its potential as a marker for MP contamination in wetland soils (Liu et al. 2023). Significant MP concentrations were reported in the Yellow

River by Liu et al. (2021), with average abundances ranging from 5358 to 654,000 n/m^3 in water and 43.57 to 615 items/kg in sediments. The findings indicated higher MP concentrations in sediment samples compared to water samples. The sources of MPs were closely associated with agricultural and industrial activities, as well as biological processes within the habitats. The identified MPs included PE, PVC, PP, PET, and PS. Fibers remained the dominant MP kind, accounting for 68.18% to 98.93% of the total MPs (Liu et al. 2023; Duan et al. 2020; Liu et al. 2021).

6.4.4 *Amazon Delta*

Recent research by dos Santos et al. (2024) indicates that MP abundance in the region varies significantly, ranging between 5 and 74,500 MPs per m^3 in water, 0 and 8178 MPs per kg in soil residues, and 0.34–38.3 MPs per individual in aquatic organisms (dos Santos et al. 2024). Notably, nanoplastics (<100 μm) have been identified in sediments. The predominant colors of MPs were blue and colorless, originating from fibers and fragments and also found in PET and PP (dos Santos et al. 2024). The abundance of MPs in Amazon water systems is more than the ones found in other water bodies such as Magdalena in Colombia, the Guayas in Ecuador, and Indonesia's Surabaya in correlation to zones with extreme anthropogenic commotion, that is, the Yellow River from China and Guanabara Bay in Brazil (Gerolin et al. 2020). Contributing factors to this pollution include inadequate sanitation infrastructure, urban planning deficiencies, waste management issues, and an extensive network of navigable waters (Galafassi et al. 2019).

6.4.5 *Nile Delta*

Shabaka (2022) suggested that urbanization and industrial activities, with significant contributions from agricultural runoff and land use changes, add around 80 to 106 billion MPs discharged into the Mediterranean Sea through Nile estuaries annually (Shabaka et al. 2022). MPs pose ecological risks in the Nile Delta estuaries, with concentrations ranging from 761 ± 319 to 1718 ± 1008 MPs/m^3 in surface waters of the Rosetta and Damietta branches. Sediment samples from these areas showed MP concentrations ranging from 167 ± 137 to 1630 ± 1303 MPs/kg. Seven different types of polymers were identified, including PP, PE, and PET, with MPs primarily comprised of tinted and shiny particles fewer than 500 μm in size. Hassan et al. (2023) revealed that mesoplastics and MPs were present in residues, water, and a few reasonably important freshwater creatures at the Nile River due to factory undertakings, town enhancement, fishing ponds, tourism, shipping, and also harbor activities (Hassan et al. 2023). The Naga Hammadi area has an excess amount of mesoplastics and MPs in animals, water, and soil samples. A study found that larger fish ingest sizable plastic fragments; consequently, the presence of MPs in the digestive systems of aquatic organisms is highly concerning (Shabaka et al. 2022; Hassan et al. 2023).

6.4.6 *Irrawaddy Delta*

Flora and Fauna International (FFI) and Thant Myanmar released a report stating that plastic waste of around 120 tons enters the lifeline of the Myanmar Irrawaddy River daily. The study exhibited that Myanmar's coastline was extremely influenced by MPs, and the pollution was up to 28,000 fragments/km^2 (Jeske 2019). A study revealed that the quantity of MPs in the water bodies is 17 times more in the monsoon season compared to summer. Myanmar's countryside zones are in danger because there is no solid waste disposal system, and most of the waste is disposed of straight into the streams (Sarah Pocock 2019).

6.4.7 *Rhine–Meuse–Scheldt Delta*

Mani et al. (2015) underscored the Rhine River's role as a major pathway for MPs to reach the ocean and studied the surface MPs over an extensive 820 km, revealing high concentrations (average 892,777 particles/km^2), particularly in urbanized areas like the Rhine-Ruhr metropolitan region, where levels peaked at 3.9 million particles/km^2, WWT plants, tributaries, and structures like weirs, which act as both sources and sinks for MPs (Mani et al. 2015). Erni-Cassola et al. (2024) carried out a comprehensive study near Basel and sampled five cross-sectional points of the Rhine River over 22 months and found a median MPS concentration of 4.48 n/m^3 for particles sized between 50 and 3000 μm. Estimated MPS load varied widely, from 4.04×10^2 to 3.57×10^5 particles per second (Erni-Cassola et al. 2024). Further, the Table 6.1 presents the concentration of MPs, their size, and the type of polymer present in different aquatic ecosystems around the world.

6.5 MPs in Indian Delta Ecosystems

Neelavannan and Sekhar Sen (2023) proclaimed that the providence, sources, and transference of MPs in the Indian sub-continent are still little-known compared to marine ecosystems (Neelavannan and Sen 2023). Larger MPs (>500 μm) were identified using the ATR-FTIR (attenuated total reflection Fourier transform infrared) spectroscopic analysis, while FTIR with a confocal microscope was used to analyze the MPs of <500 μm (Neelavannan and Sen 2023). Gupta et al. (2024) researched the degree of plastic pollution in superficial water, the water pilaster, and sediments near Ganges–Yamuna rivers, indicating the presence of MPs ranging from 300 μm to 1 mm in fiber-shaped blue and black MPs (Gupta et al. 2024). Domestic sewage discharge, throwaway plastics, industrial trash, the fishery and textile industries, tire abrasion, religious gifts, and other factors were identified as potential sources of macroplastics and MPs (Gupta et al. 2024). Kumar et al. (2024) focused on assessing MPs in soil residues of Lake Chilika, using a combination of density separation

Table 6.1 Concentration of MPs, their size, and the type of polymer present in different aquatic ecosystems around the world

S. No.	Country	Ecosystem	Conc. of MPs	Size	Type of polymers	Identification	References
1	Indonesia	River basin	53.8% of the total	1 mm	PS, silicone polymer, poly	Stereo microscope with camera and ATR-FTIR	Ismanto et al. (2023)
2	Bangladesh	Sundarbans mangrove forest	Max: quaternary consumers (4.17 items/individual), min: primary consumers (0.56 items/individual)	<5 mm	–	μ-FTIR	Sarker et al. (2022)
3	Western Cape, South Africa	Plankenburg River	Max: spring (5.13 ± 6.62 MP/L), min: autumn (1.52 ± 2.54 MP/L)	500–1000 μm	PE (57.86%), cotton (29.84%), PP (6.68%%), and PET (5.62%)	FTIR-ATR	Apetogbor et al. (2023)
4	Spain	Ebro Delta (Mediterranean Sea)	7000 ng/L	< 1000 and 2000 Da	PS, PE, PI, PBD, PP and polysiloxanes	Chromatography coupled with mass spectrometry	Llorca et al. (2021)
5	Riau	Sago River	–	5 mm	PS, nylon, cellulose acetate (CA), HDPE, PVC	FTIR	Priyambada et al. 2023)
6	Netherland	Dutch River Delta and Amsterdam canals	68–910 L^{-1} and 48–187 L^{-1} wet weight (WW),	10 and 5000 μm	–	FTIR	Leslie et al. (2017)
7	China	Yangtze River Delta	0.48 to 21.52 items L^{-1}	<0.5 mm	PES, PP	ATR and μ-FTIR	Hu et al. (2018)
8	South China	Beijing and Pearl River Delta	3183 items/m^3 to 7571 items/m^3	<1 mm	–	Stereo microscope (OLYMPUS SZX7)	Wang et al. (2020)

(continued)

Table 6.1 (continued)

S. No.	Country	Ecosystem	Conc. of MPs	Size	Type of polymers	Identification	References
9	China	Yellow River Delta	136 to 2060 items/kg	13 mm to 5 mm	PET, PC	Optical microscope	Duan et al. (2020)
10	Malaysia	Cherating River (aquatic)	0.0070 ± 0.0033 particles/m^3	0.5 to 1.0 mm	PE, PS, PET	Binocular dissection microscope	Pariatamby et al. (2020)
11	Vietnam	Peatland areas	192.3 ± 261.3 items kg − 1	300–1000 μm	PVC (46.2%), PE (20.9%), PP (9.2%)	FTIR	Nguyen et al. (2022)
12	Egypt	Nile Delta	761 ± 319 to 1718 ± 1008 MPs/m^3 (surface water)	<500 μm	PE, PP, PET, PEVA, and PTFE	–	Shabaka et al. (2022)
13	Bangladesh	Lakes in Savar, Dhaka	19 MPs/individual	1–5 mm	PA, PS, ABS, EVA	FTIR	Shetu et al. (2023)
14	China	Yangtze River Delta	0.37–8.9 particles per m^3	–	PE, PVC	Digital optical microscope, m-Raman	Xu et al. (2024)
15	Italy	Calabria (center of the Mediterranean basin)	0.13 ± 0.19 particles/m^2	50 to 5000 μm	PE, PP	ATR FTIR	Marrone et al. (2021)
16	Eastern China	Urban rivers (Nanjing)	470 ± 119.56 items per L	38–75 mm, 270–5000 mm	PVC, PC, PE	ATR-FTIR	Wang et al. (2024)
17	India	Freshwater lake system (lesser Himalayas)	02–64 fragments/L and 15–632 fragments/kg DW	25–150 μm	DBP, DEHP	–	Ajay et al. (2021)
18	Bangladesh	Salt pans (Maheshkhali Channel)	78–137 particles kg^{-1}	250–5000 μm	PP, PET, PS, PE	FT-MIR-NIR	Rakib et al. (2021)

19	India	Ganges–Brahmaputra Delta	718 ± 244 items/m^3	0.3 mm–90 μm	HDPE, POM, PPS, PA, PP, PTFE	BIOATR-FTIR	Neelavannan et al. (2023)
20	Ghana	Eastern Atewa and Weija (Accra)	Black-chinned Tilapia in Densu Delta (2.25) MP/ person and Weija Dam (2.50) MP/ person Bagrid catfish–Densu Delta (4.17) MP/ individual and the Weija Dam (1.58) MP/ individual	Water (1.208 ± 0.264) and sediment (0.267 ± 0.051)	–	Leica EZ4 HD stereo microscope with image analyses system IC80 HD camera	Blankson et al. (2022)

Note: *LDPE* low density polyethylene, *PES* polyether sulfone, *PA* polyamide, *PEVA* polyethylene-co-vinyl acetate, *Pac* polyacrylates, *PET* polyethylene terephthalate, *PC* polycarbonate, *PE* polyethylene, *PP* polypropylene, *PS* polystyrene, *PVC* polyvinyl chloride, *PU* polyether urethane, *POM* polyoxymethylene, *PPS* polyphenylene sulfide, *DBP* dibutyl phthalate, *PTFE* polytetrafluoroethylene, *DEHP* di-(2-ethylhexyl) phthalate, *WWTP* wastewater treatment plant

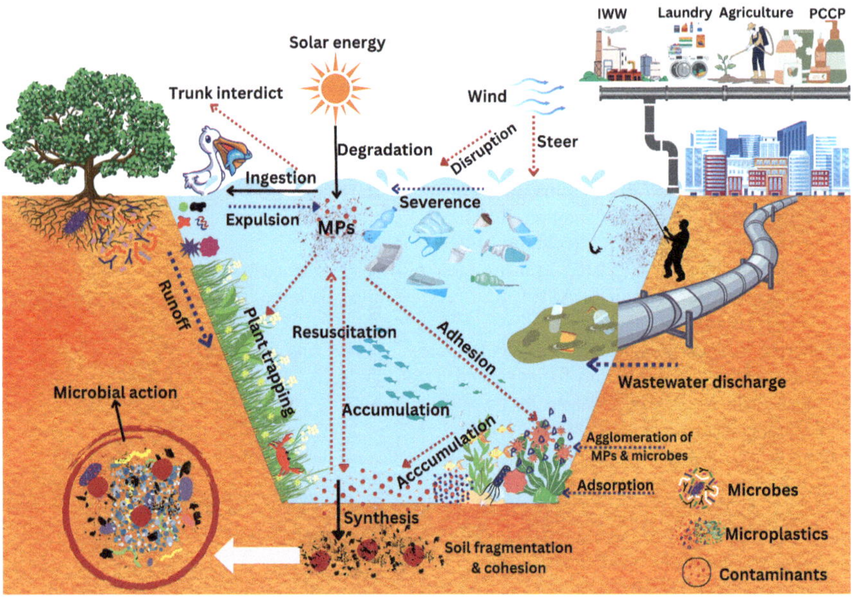

Fig. 6.4 Sources and occurrence of MPs in the delta ecosystem

and vacuum filtration methods, and found MPs at an average concentration of 440 ± 3.53 fragments per kg of samp sediment, with particles mainly reaching from 50 to 500 μm (Kumar et al. 2024). Fibers and fragments were the most common morphologies, measured at 210 ± 1.66 and 175 ± 1.76 traces per kilogram, correspondingly. A variety of polymers were identified using Raman spectroscopy, with HDPE (37%) being the most prevalent, followed by polycarbonate (7%), polypropylene (7%), polyamide (11%), polyethylene terephthalate (18%), and polystyrene (20%). MPs exhibited a range of colors, with black particles being the most common (38.8%), followed by jade, red, sapphire, white, and translucent particles. These color variations reflect diverse sources and varying degrees of environmental degradation. The primary sources of MPs were antifouling coatings, tourism, fishing activities, and aquafarming, which resulted in making Chilika Lake fall under risk category 1, indicating minimal MPs, based on the Pollution Load Index (PLI) (Kumar et al. 2024). Figure 6.4 shows the sources and occurrence of MPs in the delta ecosystem.

6.6 Impacts on the Delta Ecosystem Due to MPs

MPs are ubiquitous in nature and are measured in various size ranges, i.e., 1 mm to 10 mm (Van Cauwenberghe et al. 2015; Claessens et al. 2011), whereas nanoparticles have been documented with sizes less than 100 nm (Pramanik et al. 2021).

Consequently, the pervasive plastic waste in the ecosystem has garnered public concern about the effects of MPs on living organisms.

6.6.1 Outcomes of MP Consumption

The widespread presence of MPs in diverse organisms poses a significant threat to aquatic and terrestrial ecosystems. While numerous studies have investigated MP impacts on marine life, research on MP accumulation in terrestrial systems, particularly on soil surfaces and its ecological consequences, remains limited (Thomas et al. 2020). In aquatic environments, organisms ingest MPs through direct consumption, filter feeding, water intake, or trophic transfer from contaminated prey (Polprasert et al. 2023; Nelms et al. 2018). Filter-feeding species are especially vulnerable, exposing organisms across all trophic levels to MP contamination (Andrady 2011).

Research in 2018 investigated the ingestion of MPs by *Megaptera novaeangliae* whales in the Wadden Maritime (Besseling 2018). The gastrointestinal tract samples contained 45 substances of suspected synthetic origin. The polymer kinds discovered remained PVC, PA, PE, PET, and PP. The organism's weak capacity to distinguish between prey and plastics makes MPs easier to absorb (Gomiero et al. 2019). Furthermore, research done in the North Pacific Ocean found how maritime plankton frequently confuse gray and softly tinted minute plastic fragments as eatable items (Onyango 2020; Bai et al. 2021).

The consumed MPs are digested by the body and remain in the GI system (Kaseke et al. 2023). Browne et al. (2011) conducted an experiment to illustrate the transfer of MPs among marine organisms (Browne et al. 2011). The model organism, *Mytilus edulis*, was injected with fluorescent microspheres, revealing its capacity to absorb MPs measuring 2–4 μm; after 3 days, MPs ranging from 3 to 9.6 μm were detected in the blood. Despite the absence of obvious detrimental effects before the ingestion of MPs by *bivalve molluscs*, nodular tenderness was detected in the GI tract (Von Moos et al. 2012). Shang et al. (2020) demonstrated the adsorption of MPs by terrestrial creatures; specifically, *Caenorhabditis elegans* can ingest PS beads (Shang et al. 2020). The circulation of MPs in aquatic and terrestrial food chains forming a food web is represented in Fig. 6.5.

This nematode may serve as a possible vector for the introduction of MPs into terrestrial food webs (Wang et al. 2019). However, the effects of ingesting noncontaminated MPs have to be determined. The ingestion of MPs can lead to physical damage, toxicity, and genetic alterations in organisms. The gulped MPs may aggregate and obstruct the appendage, hindering the transit of diet hooked onto the duodenal tracts (Hou et al. 2022). This physical damage ultimately diminishes the organism's capacity to absorb nutrients (Wright et al. 2013). A researcher in 2014 documented the ingestion of MPs (PE) by *Idotea emarginata* throughout a 7-week watch period, resulting in reduced diet consumption (Hämer et al. 2014). The entry of MPs into maritime organisms resulted in their presence within the food chain. A

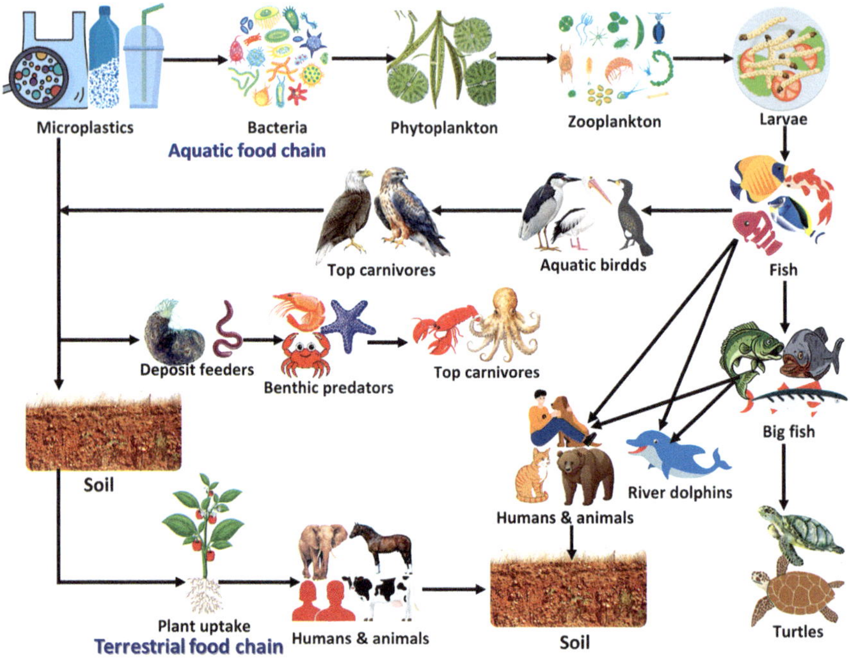

Fig. 6.5 Circulation of MPs in aquatic and terrestrial food chains forming a food web

study demonstrated the ingestion of 2 mm to 5 mm MPs by *fur seals*, presumed to be secondary polymers eaten by *Electrona subaspera* (Nelms et al. 2018). The algae exposed to nano-polystyrene led to a decrease in cellular chlorophyll-a concentration and growth, possibly due to the obstruction of nutrient flow and CO_2 (Yu et al. 2024). Prolonged exposure to MPs may result in detrimental effects on budding and parturition in an organism. While the consumption of MPs is detrimental to aquatic creatures, several living things can eliminate the plastic particles from their bodies without causing harm. *Eurytemora affinis* copepods exposed to MPs for half a day were observed to ingest and subsequently expel the particles (Coppock et al. 2019; Koski et al. 2021). Research published in 2004 also demonstrated that *polychaete* maggots could were capable of excreting ingested MPs through their fecal castings (Hodgson 2018).

6.6.2 Effects on Humans

Several years ago, when MPs were detected in the digestive systems of fish and *shellfish*, attention shifted toward the safety of seafood consumption. *Shellfish* raised particular concern because, unlike fish, humans typically consume the entire organism—including the digestive system—thereby ingesting any MPs present (Garrido

Gamarro et al. 2020). In 2017, Belgian researchers revealed that seafood enthusiasts would ingest as many as 11,000 plastic particles annually by consuming *mussels*, a popular delicacy in Belgium (Claessens et al. 2011). By that time, however, scientists had already comprehended that plastics perpetually break in the environment, gradually disintegrating into fibers thinner than human hair, particles so minuscule they readily become airborne (Carr 2017). A research team at the University of Plymouth in the U.K. aimed to evaluate the risks associated with consuming infected wild *mussels* in Scotland to those of inhaling air in a standard household (Helwig et al. 2024). Individuals are likely to absorb more plastic through the inhalation or ingestion of microscopic, imperceptible plastic fibers present in the surrounding air—released from clothing, carpets, and upholstery—than through the consumption of *mussels* (Wu et al. 2022). In 2022, scientists from the Netherlands and the U.K. reported the presence of MP particles in living humans, specifically in two previously unobserved locations: deep within the lungs of surgical patients and in the blood of anonymous donors (Leonard et al. 2024; Amato-Lourenço et al. 2021).

Maritime expeditions to quantify MPs in the ocean have produced staggering figures, which have continued to rise due to the increasing volume of plastic waste entering marine environments and breaking down over time. A 2021 estimate showed that there are 24.4 trillion MPs in the surface oceans—equivalent to about 30 billion 500-ml bottles of drinking water—a scale that is difficult to fathom (Isobe et al. 2021). A report released in 2023 said that the sum exceeded 170 trillion (Eriksen et al. 2023). MPs have infused various ecosystems, entering the human food chain via seafood, drinking water, and even air, due to their persistence and widespread use (Emenike et al. 2023). The bearings of MPs on human well-being are illustrated in Fig. 6.6.

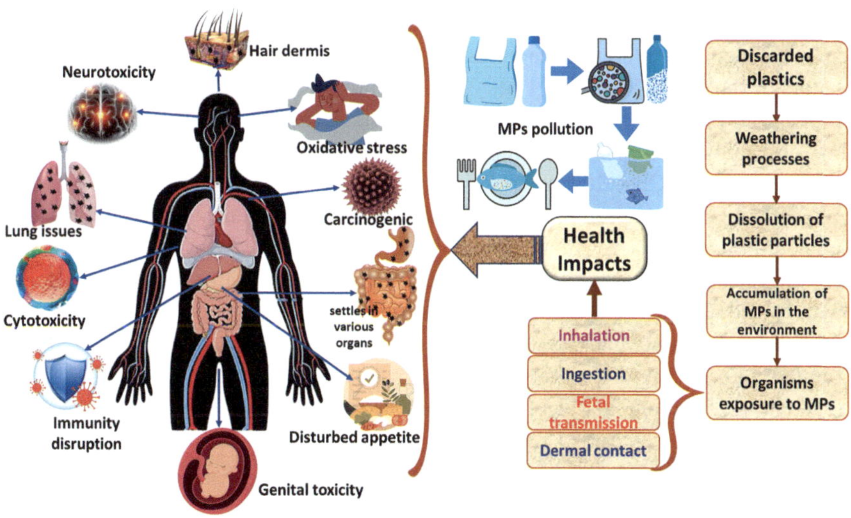

Fig. 6.6 Formation of MPs and their effects on human health through consumption

6.7 Sampling and Detection of MPs

MPs are ubiquitous pollutants found in nearly all environmental compartments. Nevertheless, understanding of fate, sources, and environmental contamination during the time course and place remains constrained because of the arduous and diverse analytical methods presently employed. This study throws light on sample gathering, parting, digestion, singling out, quantification, and the extenuation of cross-contamination (Sekar and Sundaram 2023; Bui et al. 2020; Wang et al. 2023; Martin et al. 2018). The selection of MPs would enhance representativity and reproducibility by determining density separation, filter pore size, bulk sample volume, and digestion results, as well as by employing innovative techniques, such as staining dyes for improved visual identification and the widespread application of chemical characterization (Ribeiro et al. 2024; Wang et al. 2021).

Research done by Ajao (2024) gave critical insights into the methods for detecting MPs in water samples while tackling substantial problems in quantification and standardization (Ajao 2024). Hydrogen peroxide (H_2O_2) and Fenton's reagent were recognized as efficient for oxidative digestion while minimally affecting the characteristics of MPs (Ajao 2024). A review by Rocha-Santos and Duarte (2015) provided a comprehensive analysis of MPs in the environment, focusing on their occurrence, fate, and behavior (Rocha-Santos and Duarte 2015). It discussed the pervasive nature of MPs, which are defined as plastic units lesser than 5 mm, and their significant impact on both terrestrial and marine ecosystems (Rocha-Santos and Duarte 2015). Wenfeng Wang and Jun Wang (2018) advocated for combining visual methods with spectroscopic techniques like FTIR and Raman spectroscopy for accurate identification and highlighted the urgent need for standard methodologies for detecting and analyzing MPs in environmental samples and also suggested that the efficiency of MP extraction from environmental matrices is influenced by choice of extraction solutions (Wang and Wang 2018). The use of heavy salts (e.g., $ZnCl_2$) is recommended to enhance extraction effectiveness while minimizing environmental hazards (Razeghi et al. 2022). Jieun Lee and Kyu-Jung Chae (2021) emphasized the urgent need to establish systematic protocols for sampling, extraction, and analysis to ensure accuracy, reliability, and reproducibility in microplastic research (Lee and Chae 2021). Apart from FTIR and Raman spectroscopy, differential scanning calorimetry (DSC) and thermogravimetric analysis (TGA) have also been employed to investigate the thermal degradation pathways of MPs, their endothermic phase transitions, melting characteristics, and reaction kinetics (Sorolla-Rosario et al. 2022; Dang et al. 2024). X-ray diffraction helps in finding the structural and functional group analysis and degree of crystallinity (Moura et al. 2023). SEM coupled with energy dispersive spectroscopy (SEM/EDS) helped in finding structural characterization and elemental composition in MPs (Wang et al. 2017). Figure 6.7 gives a generalized illustration of the steps involved in MP sampling and characterization.

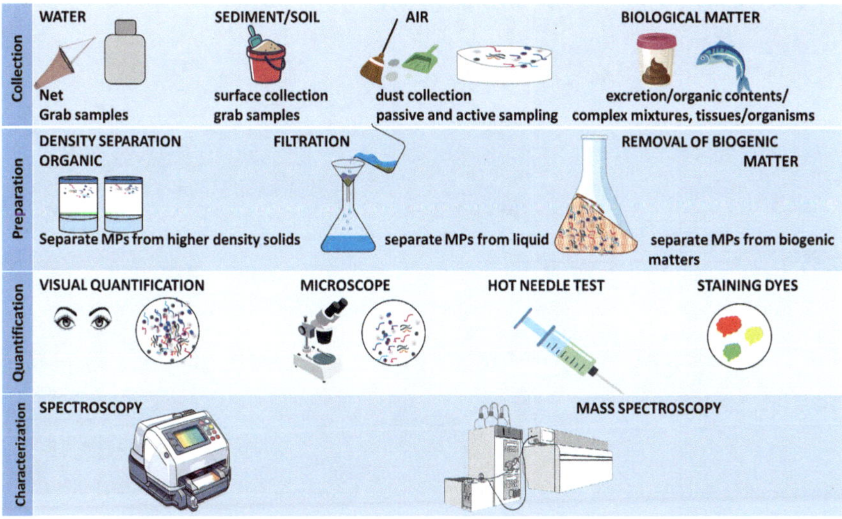

Fig. 6.7 Steps involved in MP sampling and characterization

6.8 Restoring Delta Ecosystems

The delicate ecosystems of deltas face a threat from pollution that can harm the variety of species and services they provide to the environment and communities living around them. Safeguarding these ecosystems effectively and sustainably requires an approach that includes preventive measures to stop plastic waste from entering rivers and seas, as well as reducing the use of disposable plastics through community participation and implementing regulations. Furthermore, regular cleaning efforts and the deployment of collection devices can also play a key role in mitigating the impact of contamination in these crucial ecosystems. Figure 6.8 shows the effort adopted for the Eluru canal near Vijayawada railway station by the Andhra Pradesh government as part of the Mission for Clean Krishna and Godavari (MCKGC). The image also shows the status of plastic pollution at Monga River near the Sundarbans, an irrigation canal near Vadlamudi, Guntur, and Eluru canals. Engagement and involvement of the community play a role in promoting practices through awareness campaigns and local empowerment initiatives that encourage people to adopt eco-friendly behaviors. Efforts to address issues include banning plastics and promoting eco-labeled products through policy measures and international collaboration. Scientific studies and monitoring programs are essential for evaluating pollution levels and exploring alternatives to combat degradation. Actions such as restoring habitats like mangroves and protecting buffer zones can have impacts on preserving ecosystems for generations. Principles of eco-tourism and sustainable fishing methods play a role in encouraging tourism and fishing

Fig. 6.8 Plastic pollution at (**a**) the Monga River (PC: Asha Zahid 2023), (**b**) the Vadlamudi irrigation canal, (**c**) the Eluru Canal (PC: KVS Giri 2022), and (**d**) floating waste collection at Eluru canal gates

practices to protect delta ecosystems from the harmful effects of plastic pollution while preserving their biodiversity and ecosystem services with the help of businesses and nonprofit organizations working together to support these efforts. With the help of artificial intelligence and machine learning, it is possible to categorize the varieties of MPs present in the delta's main aquatic environments and determine their concentrations.

6.9 Conclusions

Delta ecosystems are very significant around the world for the security of food, diversity of life, and maintenance of numerous living beings; however, they come with serious fears from MP pollution. Research related to crucial deltas, such as Hooghly, Nile, Sundarbans, etc., exhibit MPs in residues, water, and aquatic creatures in high percentages compared to other aquatic structures. These cheap strands, fragments, and films cross the threshold of the basic cell level, destroying tissues, poisoning the food hierarchy, and disturbing reproduction. MPs are divided equally through water and sediment, making their containment difficult.

The primary causes of pollution are town wastes, industrial trash, plastic trash from tourism, and farming runoff, which aggravates the number of MPs and pushes them straight into food stock. Delta areas are important farmlands and fish-catching areas, so MP pollution risks food certainty, water quality standards, and public well-being. Safeguarding ecological systems requires quick measures such as government intervention to control pollution. A study needs to be undertaken on MPs and their impacts. People's realization and community measures become crucial in decreasing pollution and protecting the delta ecosystems for generations to come.

References

Ajao A (2024) Conventional and advanced treatment technologies for Microplastics in water treatment facilities. Preprints. https://doi.org/10.20944/preprints202406.0075.v1

Ajay K, Behera D, Bhattacharya S, Mishra PK, Ankit Y, Anoop A (2021) Distribution and characteristics of microplastics and phthalate esters from a freshwater lake system in lesser Himalayas. Chemosphere 283:131132

Allison MA (1998) Historical changes in the Ganges-Brahmaputra Delta front. Autumn

Amato-Lourenço LF, Carvalho-Oliveira R, Júnior GR, dos Santos GL, Ando RA, Mauad T (2021) Presence of airborne microplastics in human lung tissue. J Hazard Mater 416:126124

Andrady AL (2011) Microplastics in the marine environment. Mar Pollut Bull 62:1596–1605

Apetogbor K, Pereao O, Sparks C, Opeolu B (2023) Spatio-temporal distribution of microplastics in water and sediment samples of the Plankenburg river, Western Cape, South Africa. Environ Pollut 323:121303

Bai Z, Wang N, Wang M (2021) Effects of microplastics on marine copepods. Ecotoxicol Environ Saf 217:112243

Besseling E (2018) Micro- and nanoplastic in the aquatic environment: from rivers to whales. Wageningen University

Blankson ER, Tetteh PN, Oppong P, Gbogbo F (2022) Microplastics prevalence in water, sediment and two economically important species of fish in an urban riverine system in Ghana. PLoS One 17:e0263196. https://doi.org/10.1371/journal.pone.0263196

Borriello L, Scivicco M, Cacciola NA, Esposito F, Severino L, Cirillo T (2012) Microplastics, a Global Issue: Human Exposure through Environmental and Dietary Sources. Food 12(18):3396

Browne MA, Crump P, Niven SJ, Teuten E, Tonkin A, Galloway T, Thompson R (2011) Accumulation of microplastic on shorelines woldwide: sources and sinks. Environ Sci Technol 45:9175–9179

Bui XT, Vo TDH, Nguyen PT, Nguyen VT, Dao TS, Nguyen PD (2020) Microplastics pollution in wastewater: characteristics, occurrence and removal technologies. Environ Technol Innov 19:101013

Carr SA (2017) Sources and dispersive modes of micro-fibers in the environment. Integr Environ Assess Manag 13:466–469

Cera A, Gallitelli L, Scalici M (2023) Macroplastics in lakes: an underrepresented ecological problem? Water (Switzerland) 15:60

Chen S, Li Y, Mawhorter C, Legoski S (2021) Quantification of microplastics by count, size and morphology in beverage containers using Nile Red and ImageJ. J Water Health 19:79–88

Claessens M, De MS, Van LL, De CK, Janssen CR (2011) Occurrence and distribution of microplastics in marine sediments along the Belgian coast. Mar Pollut Bull 62:2199–2204

Coppock RL, Galloway TS, Cole M, Fileman ES, Queirós AM, Lindeque PK (2019) Microplastics alter feeding selectivity and faecal density in the copepod, Calanus helgolandicus. Sci Total Environ 687:780–789

Dang TT, Sogut E, Unalan IU, Corredig M (2024) Quantification of polystyrene microplastics in water, milk, and coffee using 2 thermogravimetry coupled with fourier transform infrared spectroscopy (TGA-3 FTIR) 4

dos Santos SJ, Cidade MJA, Panero F d S, Ribeiro LB, Campos da Rocha FO (2024) Microplastic pollution in the Amazon Basin: current scenario, advances and perspectives. Sci Total Environ 946:174150

Dris R, Gasperi J, Mirande C, Mandin C, Guerrouache M, Langlois V, Tassin B (2017) A first overview of textile fibers, including microplastics, in indoor and outdoor environments. Environ Pollut 221:453–458

Duan Z, Zhao S, Zhao L, Duan X, Xie S, Zhang H, Liu Y, Peng Y, Liu C, Wang L (2020) Microplastics in Yellow River Delta wetland: occurrence, characteristics, human influences, and marker. Environ Pollut 258:113232

Emenike EC, Okorie CJ, Ojeyemi T, Egbemhenghe A, Iwuozor KO, Saliu OD, Okoro HK, Adeniyi AG (2023) From oceans to dinner plates: the impact of microplastics on human health. Heliyon 9:e20440

Eriksen M, Cowger W, Erdle LM, Coffin S, Villarrubia-Gómez P, Moore CJ, Carpenter EJ, Day RH, Thiel M, Wilcox C (2023) A growing plastic smog, now estimated to be over 170 trillion plastic particles afloat in the world's oceans—urgent solutions required. PLoS One 18:e0281596

Erni-Cassola G, Dolf R, Burkhardt-Holm P (2024) Microplastics in the water column of the Rhine River near Basel: 22 months of sampling. Environ Sci Technol 58:5491–5499

Galafassi S, Nizzetto L, Volta P (2019) Plastic sources: a survey across scientific and grey literature for their inventory and relative contribution to microplastics pollution in natural environments, with an emphasis on surface water. Sci Total Environ 693:133499

Gao Y, Fan K, Wang C, Zeng Y, Li H, Mai Y, Liu Q, Lai Z (2022) Abundance, composition, and potential ecological risks of microplastics in surface water at different seasons in the Pearl River Delta, China, vol 14, Water (Switzerland), p 2545

Garrido Gamarro E, Ryder J, Elvevoll EO, Olsen RL (2020) Microplastics in fish and shellfish–a threat to seafood safety? J Aquat Food Prod Technol 29:417–425

Gerolin CR, Pupim FN, Sawakuchi AO, Grohmann CH, Labuto G, Semensatto D (2020) Microplastics in sediments from Amazon rivers, Brazil. Sci Total Environ 749:141604

Gomiero A, Strafella P, Fabi G (2019) From macroplastic to microplastic litter: occurrence, composition, source identification and interaction with aquatic organisms. Experiences from the Adriatic Sea. In: Plastics in the environment. IntechOpen

Gupta P, Saha M, Naik A, Kumar MM, Rathore C, Vashishth S, Maitra SP, Bhardwaj KD, Thukral H (2024) A comprehensive assessment of macro and microplastics from Rivers Ganga and Yamuna: unveiling the seasonal, spatial and risk factors. J Hazard Mater 469:133926

Hämer J, Gutow L, Köhler A, Saborowski R (2014) Fate of microplastics in the marine isopod idotea emarginata. Environ Sci Technol 48:13451–13458

Hassan YAM, Badrey AEA, Osman AGM, Mahdy A (2023) Occurrence and distribution of meso- and macroplastics in the water, sediment, and fauna of the Nile River, Egypt. Environ Monit Assess 195:1130

He S, Chi HY, Li C, Gao Y, Li ZC, Zhou XX, Yan B (2022) Distribution, bioaccumulation, and trophic transfer of palladium-doped nanoplastics in a constructed freshwater ecosystem. Environ Sci Nano 9:1353–1363

Helwig K, Cooper A, Pagaling E, Henderson F, Avery L, Roberts J et al (2024) Emerging contaminants: informing scotland's strategic monitoring and policy approaches on substances of increasing concern. Centre of Expertise for Waters (CREW). https://www.crew.ac.uk/sites/www.crew.ac.uk/files/publication/CRW2022_06_Emerging%20contaminants%20report%20and%20appendices_0.pdf

Hoang TTT, Nguyen BA, Pham NNQ, Nguyen NB, Tran TKT, Tu TCL, Huynh TD, Nguyen TNT, Pham TPQ, Nguyen TV, Nguyen TL, Le QH (2022) The potential emission of personal care products derived plastic microbeads: a case study of Ho Chi Minh City, Vietnam. In: IOP conference series: earth and environmental science. IOP Publishing Ltd

Hodgson D (2018) The impacts of microplastic ingestion on marine polychaete worms. University of Exeter (United Kingdom), ProQuest Dissertations & Theses, 2018. 30397810

Hou M, Xu C, Zou X, Xia Z, Su L, Qiu N, Cai L, Yu F, Wang Q, Zhao X, He Y, Wang C, Wang J (2022) Long-term exposure to microplastics induces intestinal function dysbiosis in rare minnow (Gobiocypris rarus). Ecotoxicol Environ Saf 246:114157

Hu L, Chernick M, Hinton DE, Shi H (2018) Microplastics in small waterbodies and tadpoles from Yangtze River Delta, China. Environ Sci Technol 52:8885–8893

Islam SN, Gnauck A (2008) Mangrove wetland ecosystems in Ganges-Brahmaputra delta in Bangladesh. Front Earth Sci China 2:439–448

Ismanto A, Hadibarata T, Sugianto DN, Zainuri M, Kristanti RA, Wisha UJ, Hernawan U, Anindita MA, Gonsilou AP, Elshikh MS, Al-Mohaimeed AM, Abbasi AM (2023) First evidence of microplastics in the water and sediment of Surakarta city river basin, Indonesia. Mar Pollut Bull 196:115677

Isobe A, Azuma T, Cordova MR, Cózar A, Galgani F, Hagita R, Kanhai LD, Imai K, Iwasaki S, Kako S, Kozlovskii N, Lusher AL, Mason SA, Michida Y, Mituhasi T, Morii Y, Mukai T, Popova A, Shimizu K, Tokai T, Uchida K, Yagi M, Zhang W (2021) A multilevel dataset of microplastic abundance in the world's upper ocean and the Laurentian Great Lakes. Microplast Nanoplast 1:16

Jeske F (2019) Survey on Plastic Waste in the Ayeyarwady, 2018-2019: Rapid river sampling for the first quantitative assessment of floating plastics in Myanmar's great river. Working Paper No.09 of FFI Myanmar, with Thant Myanmar

Kalčíková G, Alič B, Skalar T, Bundschuh M, Gotvajn AŽ (2017) Wastewater treatment plant effluents as source of cosmetic polyethylene microbeads to freshwater. Chemosphere 188:25–31

Kankanige D, Babel S (2021) Contamination by ≥6.5 µm-sized microplastics and their removability in a conventional water treatment plant (WTP) in Thailand. J Water Process Eng 40:101765

Kaseke T, Lujic T, Cirkovic Velickovic T (2023) Nano- and microplastics migration from plastic food packaging into dairy products: impact on nutrient digestion, absorption, and metabolism. Food Secur 12:3043

Kim S, Hyeon Y, Rho H, Park C (2024) Ceramic membranes as a potential high-performance alternative to microplastic filters for household washing machines. Sep Purif Technol 344:127278

Klein S, Dimzon IK, Eubeler J, Knepper TP (2018) Analysis, occurrence, and degradation of microplastics in the aqueous environment. In: Handbook of environmental chemistry. Springer, pp 51–67

Koski M, Søndergaard J, Christensen AM, Nielsen TG (2021) Effect of environmentally relevant concentrations of potentially toxic microplastic on coastal copepods. Aquat Toxicol 230:105713

Kuenzer C, Renaud FG (2012) Climate and environmental change in river deltas globally: expected impacts, resilience, and adaptation, pp 7–46

Kumar R, Sharma P, Bandyopadhyay S (2021) Evidence of microplastics in wetlands: extraction and quantification in freshwater and coastal ecosystems. Journal of water. Process Eng 40:101966

Kumar R, Sinha R, Refat Jahan Rakib M, Padha S, Ivy N, Bhattacharya S, Dhar A, Sharma P (2022) Microplastics pollution load in Sundarban delta of Bay of Bengal. J Hazard Mater Adv 7:100099

Kumar M, Naik DK, Maharana D, Das M, Jaiswal E, Naik AS, Kumari N (2024) Sediment-associated microplastics in Chilika lake, India: highlighting their prevalence, polymer types, possible sources, and ecological risks. Sci Total Environ 914:169707

Lee J, Chae KJ (2021) A systematic protocol of microplastics analysis from their identification to quantification in water environment: a comprehensive review. J Hazard Mater 403:124049

Leonard VL, Liddle S CR, Atherall CA, Chapman E, Watkins M, Calaminus DJS, Rotchell JM (2024) Microplastics in human blood: polymer types, concentrations and characterisation using µFTIR. Environ Int 188:108751

Leslie HA, Brandsma SH, van Velzen MJM, Vethaak AD (2017) Microplastics en route: field measurements in the Dutch river delta and Amsterdam canals, wastewater treatment plants, North Sea sediments and biota. Environ Int 101:133–142

Lin YD, Huang PH, Chen YW, Hsieh CW, Tain YL, Lee BH, Hou CY, Shih MK (2023) Sources, degradation, ingestion and effects of microplastics on humans: a review. Toxics 11:747
Liu R-p, Li Z-z, Liu F, Dong Y, Jiao J-g, Sun P-p, El-Wardany RM (2021) Microplastic pollution in Yellow River, China: current status and research progress of biotoxicological effects. China Geol 4:585–592
Liu S, You H, Mu H, Cheng J, Kuang S, Wang F, Chen H, Zheng M, Xu Y, Liu T (2023) Abundance, characteristics and risk assessment of microplastics in aquatic sediments: a comparative study in the Yellow River and Yellow Sea. Waste Manag 172:326–334
Llorca M, Vega-Herrera A, Schirinzi G, Savva K, Abad E, Farré M (2021) Screening of suspected micro(nano)plastics in the Ebro Delta (Mediterranean Sea). J Hazard Mater 404:124022
Mani T, Hauk A, Walter U, Burkhardt-Holm P (2015) Microplastics profile along the Rhine River. Sci Rep 5:17988
Marrone A, La Russa MF, Randazzo L, La Russa D, Cellini E, Pellegrino D (2021) Microplastics in the center of mediterranean: comparison of the two calabrian coasts and distribution from coastal areas to the open sea. Int J Environ Res Public Health 18:10712
Martin KM, Hasenmueller EA, White JR, Chambers LG, Conkle JL (2018) Sampling, sorting, and characterizing microplastics in aquatic environments with high suspended sediment loads and large floating debris. J Vis Exp 28:57969
Moura DS, Pestana CJ, Moffat CF, Hui J, Irvine JTS, Lawton LA (2023) Characterisation of microplastics is key for reliable data interpretation. Chemosphere 331:138691
Nawar N, Rahman MM, Chowdhury FN, Marzia S, Ali MM, Akbor MA, Siddique MAB, Khatun MA, Shahjalal M, Huque R, Malafaia G (2023) Characterization of microplastic pollution in the Pasur river of the Sundarbans ecosystem (Bangladesh) with emphasis on water, sediments, and fish. Sci Total Environ 868:161704
Neelavannan K, Sen IS (2023) Microplastics in freshwater ecosystems of India: current trends and future perspectives. ACS Omega 8:34235–34248
Neelavannan K, Sen IS, Sinha N, Thakur AK, Misra S (2023) Microplastics in the Ganga-Brahmaputra delta: sources and pathways to the Sundarbans biosphere reserve – an UNESCO World Heritage Centre. Environ Adv 11:100350
Nelms SE, Galloway TS, Godley BJ, Jarvis DS, Lindeque PK (2018) Investigating microplastic trophic transfer in marine top predators. Environ Pollut 238:999–1007
Nguyen MK, Lin C, Hung NTQ, Vo DVN, Nguyen KN, Thuy BTP, Hoang HG, Tran HT (2022) Occurrence and distribution of microplastics in peatland areas: a case study in long an province of the Mekong Delta, Vietnam. Sci Total Environ 844:157066
Onyango WA (2020) Types and abundance of microplastics in macro-invertebrates along the Kenyan Coast. Afr J Mar Sci 42(4):405–411
Oßmann BE, Sarau G, Holtmannspötter H, Pischetsrieder M, Christiansen SH, Dicke W (2018) Small-sized microplastics and pigmented particles in bottled mineral water. Water Res 141:307–316
Pariatamby A, Hamid FS, Bhatti MS, Anuar N, Anuar N (2020) Status of microplastic pollution in aquatic ecosystem with a case study on cherating river, Malaysia. J Eng Technol Sci 52:222–241
Parvin F, Jannat S, Tareq SM (2021) Abundance, characteristics and variation of microplastics in different freshwater fish species from Bangladesh. Sci Total Environ 784:147137
Perren W, Wojtasik A, Cai Q (2018) Removal of microbeads from wastewater using electrocoagulation. ACS Omega 3:3357–3364
Polprasert C, Yeemin T, Sutthacheep M, Pussayanavin T, Jinda K, Kamngam S (2023) Marine ecosystems and emerging plastic pollution. In Koottatep T, Winijkul E, Xue W, Panuvatvanich A, Visvanathan C, Pussayanavin T, Limphitakphong N, Polprasert C (eds.), Marine Plastics Abatement: Challenges, Implications, Assessments and Circularity (Vol 1). https://doi.org/10.2166/9781789063202_0085
Pramanik BK, Pramanik SK, Monira S (2021) Understanding the fragmentation of microplastics into nano-plastics and removal of nano/microplastics from wastewater using membrane, air flotation and nano-ferrofluid processes. Chemosphere 282:131053

Priyambada G, Kurniawan B, Sitompul RG, Darmayanti L (2023) The abundance of microplastics in Siak tributary sediments in the watershed area, Pekanbaru City, Riau (Case Study Sago River). In: Materials today: proceedings. Elsevier Ltd, pp 272–277
Rakib MRJ, Al Nahian S, Alfonso MB, Khandaker MU, Enyoh CE, Hamid FS, Alsubaie A, Almalki ASA, Bradley DA, Mohafez H, Islam MA (2021) Microplastics pollution in salt pans from the Maheshkhali Channel, Bangladesh. Sci Rep 11:23187
Ramesh K, Chellam PV, Sundaram B (2024) Transport of layered and spherical microplastics in aqueous ecosystems: a review. Environ Chem Lett 22:1221–1255
Razeghi N, Hamidian AH, Mirzajani A, Abbasi S, Wu C, Zhang Y, Yang M (2022) Sample preparation methods for the analysis of microplastics in freshwater ecosystems: a review. Environ Chem Lett 20:417–443
Ribeiro F, Duarte AC, da Costa JP (2024) Staining methodologies for microplastics screening. TrAC Trends Anal Chem 172:117555
Rocha-Santos T, Duarte AC (2015) A critical overview of the analytical approaches to the occurrence, the fate and the behavior of microplastics in the environment. TrAC Trends Anal Chem 65:47–53
Sarah Pocock (2019) Plastic pollution piles the pressure on Myanmar's troubled turtles by Sarah Pocock
Sarker S, Huda ANMS, Niloy MNH, Chowdhury GW (2022) Trophic transfer of microplastics in the aquatic ecosystem of Sundarbans mangrove forest, Bangladesh. Sci Total Environ 838:155896
Schneider P, Asch F (2020) Rice production and food security in Asian mega deltas—a review on characteristics, vulnerabilities and agricultural adaptation options to cope with climate change. J Agron Crop Sci 206:491–503
Scown MW, Dunn FE, Dekker SC, van Vuuren DP, Karabil S, Sutanudjaja EH, Santos MJ, Minderhoud PSJ, Garmestani AS, Middelkoop H (2023) Global change scenarios in coastal river deltas and their sustainable development implications. Glob Environ Chang 82:102736
Sekar V, Sundaram B (2023) Occurrence, quantification and characterisation of microplastics in Godavari River, India. Case Stud Chem Environ Eng 8:100542
Shabaka S, Moawad MN, Ibrahim MIA, El-Sayed AAM, Ghobashy MM, Hamouda AZ, El-Alfy MA, Darwish DH, Youssef NAE (2022) Prevalence and risk assessment of microplastics in the Nile Delta estuaries: "The Plastic Nile" revisited. Sci Total Environ 852:158446
Shang X, Lu J, Feng C, Ying Y, He Y, Fang S, Lin Y, Dahlgren R, Ju J (2020) Microplastic (1 and 5 μm) exposure disturbs lifespan and intestine function in the nematode Caenorhabditis elegans. Sci Total Environ 705:135837
Shetu MH, Parvin F, Tareq SM (2023) Identifying the presence of microplastics in frogs from the largest delta of the world. Environmental. Advances 11:100355
Sigler M (2014) The effects of plastic pollution on aquatic wildlife: current situations and future solutions. Water Air Soil Pollut 225:1
Simon-Sánchez L, Grelaud M, Garcia-Orellana J, Ziveri P (2019) River deltas as hotspots of microplastic accumulation: the case study of the Ebro River (NW Mediterranean). Sci Total Environ 687:1186–1196
Sorolla-Rosario D, Llorca-Porcel J, Pérez-Martínez M, Lozano-Castelló D, Bueno-López A (2022) Study of microplastics with semicrystalline and amorphous structure identification by TGA and DSC. J Environ Chem Eng 10:106886
Syvitski JPM (2008) Deltas at risk, vol 3. Sustain Sci, In, pp 23–32
Syvitski J, Anthony E, Saito Y, Zăinescu F, Day J, Bhattacharya JP, Giosan L (2022) Large deltas, small deltas: toward a more rigorous understanding of coastal marine deltas. Glob Planet Chang 218:103958
Thomas D, Schütze B, Heinze WM, Steinmetz Z (2020) Sample preparation techniques for the analysis of microplastics in soil—a review. Sustainability (Switzerland) 12:1–28
Van Cauwenberghe L, Claessens M, Vandegehuchte MB, Janssen CR (2015) Microplastics are taken up by mussels (Mytilus edulis) and lugworms (Arenicola marina) living in natural habitats. Environ Pollut 199:10–17

Vivekanand AC, Mohapatra S, Tyagi VK (2021) Microplastics in aquatic environment: challenges and perspectives. Chemosphere 282:131151

Von Moos N, Burkhardt-Holm P, Köhler A (2012) Uptake and effects of microplastics on cells and tissue of the blue mussel Mytilus edulis L. after an experimental exposure. Environ Sci Technol 46:11327–11335

Wang W, Wang J (2018) Investigation of microplastics in aquatic environments: an overview of the methods used, from field sampling to laboratory analysis. TrAC Trends Anal Chem 108:195–202

Wang ZM, Wagner J, Ghosal S, Bedi G, Wall S (2017) SEM/EDS and optical microscopy analyses of microplastics in ocean trawl and fish guts. Sci Total Environ 603–604:616–626

Wang W, Gao H, Jin S, Li R, Na G (2019) The ecotoxicological effects of microplastics on aquatic food web, from primary producer to human: a review. Ecotoxicol Environ Saf 173:110–117

Wang S, Zhang C, Pan Z, Sun D, Zhou A, Xie S, Wang J, Zou J (2020) Microplastics in wild freshwater fish of different feeding habits from Beijiang and Pearl River Delta regions, South China. Chemosphere 258:127345

Wang C, Jiang L, Liu R, He M, Cui X, Wang C (2021) Comprehensive assessment of factors influencing Nile red staining: eliciting solutions for efficient microplastics analysis. Mar Pollut Bull 171:112698

Wang L, Zhang J, Huang W, He Y (2023) Laboratory simulated aging methods, mechanisms and characteristic changes of microplastics: a review. Chemosphere 315:137744

Wang L, Huang J, Chen M, Jin H, Wu Y, Chen X (2024) Investigation of microplastics in urban rivers of eastern China in summer: abundance, characteristics and ecological risk assessment. Environ Sci Process Impacts 26:1245–1256

Wright SL, Thompson RC, Galloway TS (2013) The physical impacts of microplastics on marine organisms: a review. Environ Pollut 178:483–492

Wu P, Lin S, Cao G, Wu J, Jin H, Wang C, Wong MH, Yang Z, Cai Z (2022) Absorption, distribution, metabolism, excretion and toxicity of microplastics in the human body and health implications. J Hazard Mater 437:129361

Xu L, Li J, Yang S, Li Z, Liu Y, Zhao Y, Liu D, Targino AC, Zheng Z, Yu M, Xu P, Sun Y, Li W (2024) Characterization of atmospheric microplastics in Hangzhou, a megacity of the Yangtze river delta, China. Environ Sci Atmos 4:1161

Yu T, Huang X, Li H, Zheng J, Gao L, Wang S, Zhang Y (2024) Silicate derived from phaeodactylum tricornutum for removal of polystyrene: interfacial effects of living organism and its derivatives with nanoplastics. Langmuir 40:22931

Chapter 7
Saline Intrusion Dynamics and Groundwater Quality Monitoring in the Coastal Mangalore Delta via Visual MODFLOW Flex

A. R. Diana, Vijay Suryawanshi, H. Ramesh, and M. S. Ganesh Prasad

Abstract The Mangalore Delta in India exemplifies the intricacies of salinization in delta regions, influenced by coastal, fluvial, and geomorphological factors, and intensified by human activities. The global surge in groundwater extraction due to population growth and industrialization has led to significant seawater intrusion into coastal aquifers, threatening groundwater quality and posing health risks. This chapter summarizes a study investigating saline water infiltration into freshwater aquifers in the Mangalore Delta using Visual MODFLOW Flex, aiming to identify intrusion patterns and assess groundwater vulnerability. The study also examines key processes affecting groundwater quality in relation to hydrogeochemical sources in the shallow coastal aquifer. Data from 27 wells were collected, analyzing 18 physicochemical properties of groundwater including temperature, pH, oxidation-reduction potential (ORP), dissolved oxygen (DO), electrical conductivity (EC), total dissolved solids (TDS), salinity, magnesium (Mg), calcium (Ca), total hardness (TH), chlorine (Cl), sodium (Na), potassium (K), carbonate (CO_3), bicarbonate (HCO_3), and sulfate (SO_4). Contour line differences highlighted intrusion levels within the coastal zone and estuaries. The infiltration levels in the coastal zone and along the estuaries of the research area were delineated using the variation in contours. The research also investigated groundwater quality, coastal aquifer properties, model implementation, calibration, validation, and future projections.

Keywords Saline intrusion · Coastal aquifer · Groundwater quality · Visual MODFLOW flex

---

A. R. Diana · V. Suryawanshi (✉) · H. Ramesh
Department of Water Resources and Ocean Engineering, NITK, Surathkal, Karnataka, India

M. S. G. Prasad
Department of Hydraulics Engineering, NIE College, Mysore, Karnataka, India

T. V. Nagaraju, G. S. Bala (eds.), *Delta Ecosystems*, Environmental Contamination Remediation and Management, https://doi.org/10.1007/978-3-031-97920-0_7

7.1 Introduction

Coastal aquifers are crucial water sources that provide for over a billion people globally and link oceanic and hydrologic systems worldwide (Trenberth et al. 2007). The hydrogeological features of these aquifers are shaped by geological environments, mixing zones, sea-level fluctuations, and salinity-induced density gradients (Gottumukkala et al. 2024). The increasing extraction of fresh groundwater in coastal areas disrupts the existing equilibrium between freshwater and saline ocean water (Mao et al. 2006). Numerous coastal regions worldwide face severe saltwater intrusion in aquifers due to both natural and human-induced factors (Hussain et al. 2019). Excessive groundwater extraction in coastal aquifers can lead to saltwater encroachment (Abd-Elaty and Zelenakova 2022). Overall, the contamination of coastal aquifers can have significant environmental, ecological, and economic repercussions in affected areas (Roy and Datta 2017). The salinization of groundwater and the abandonment of wells near coastlines are clear indicators of seawater intrusion caused by excessive groundwater pumping (Oude Essink et al. 2010). To assess potential remediation approaches, these challenges are typically evaluated using numerical models. Effective strategies can only be developed if the physical processes in coastal aquifers are accurately simulated and examined (Hariharan and Uma Shankar 2017). Once calibrated and validated for a specific study area, mathematical models can be employed to simulate the aquifer system's response to various pumping and recharge scenarios (Zheng et al. 2012).

Various tools and techniques, such as statistical analysis, water quality indices, hydrochemical analysis, and hydrological modeling, are used to assess and monitor coastal aquifer water resources. Groundwater models, which provide a conceptual description of physical systems through mathematical equations (Oude Essink 2001), play a vital role in this process. Groundwater modeling is typically based on three primary techniques: the finite element method, the finite difference method, and the analytical element method (Berehanu et al. 2017). By simulating these processes with mathematical models, it is possible to predict and control the future distribution of saltwater in coastal aquifers. Accurate field data are essential for effective modeling and prediction. However, even when field data are limited, attempting to model the system can be informative by identifying key data gaps that are crucial for the model's success (Datta et al. 2009). This approach can help guide targeted data collection efforts. A range of models, both analytical and numerical, have been employed to study groundwater flow systems (Kaur and Kaur 2019). These models rely on sets of differential equations governing groundwater flow, with numerical methods often used to solve the equations for groundwater and contaminant movement (Sharan et al. 2024). For example, a two-dimensional finite element dispersion (2D-FED) model has been applied in coastal aquifers, while a variable-density groundwater flow and salt transport model was developed to investigate seawater intrusion in the Gulf Coast aquifers of Alabama, USA (Elango 2006). Groundwater modeling tools like MODFLOW, FEFLOW, and SEAWAT are invaluable for identifying critical depletion and recharge zones, as well as predicting future variability in these areas (Scanlon et al. 2003). These models also help evaluate the relative importance of different factors affecting recharge, provided they accurately incorporate all relevant processes (Mastrocicco et al. 2019).

While significant progress has been made in understanding the mechanisms and impacts of seawater intrusion in coastal aquifers, a critical gap remains in the comprehensive study of how long-term climate changes affect the dynamics of this intrusion. The Mangalore region heavily relies on groundwater from coastal aquifers for agriculture, industry, domestic, and aquaculture uses, intensifying the urgency to address this issue. This research seeks to fill that gap by integrating climate data, hydrogeological modeling, and predictive simulations to assess the evolving dynamics of seawater intrusion under changing climate conditions. To better understand groundwater flow and the behavior of the shallow aquifer in this region, we employ a conceptual modeling approach based on water flux calculations of hydrological components. The study utilizes Visual MODFLOW Flex v.6.1, leveraging its flow and transport engines to simulate seawater intrusion in the coastal aquifer with hypothetical values. MODFLOW software packages are chosen for their flexible modular units, which effectively represent the region's complex hydrogeological conditions. The simulation model is calibrated using observed groundwater levels, salinity data, and estimated withdrawal rates. Once calibrated, the model is applied to evaluate the impact of various pumping strategies aimed at controlling saltwater intrusion in the Mangalore coastal aquifers.

7.2 Description of the Study Area

In this chapter, the investigation was carried out on the Mangalore region's coastal strip, which is located on India's Karavali coast. Mangalore is the administrative center of the Dakshina Kannada district in Karnataka, as well as a center for business, industry, education, healthcare, and start-ups (Suryawanshi et al. 2024). The area on India's western coast is located at 12°52.2′ N 74°52.8′ E. The coastal belt covers an area of approximately 167 km^2 and has a coastline of 62 km, located between latitudes 12°46.2′ and 13°5.4′ N and longitudes 74°51′ and 74°46.2′ E (Fig. 7.1).

To track changes in the aquifers of the Mangalore coastal belt, data such as geomorphic characteristics, hydrology, and hydrogeology of the area are considered. The region is dominated by major industries such as Mangalore Refinery and Petrochemical Ltd. (MRPL), Mangalore Chemicals and Fertilizers Ltd. (MCF), and New Mangalore Port Trust (NMPT), along with numerous smaller units including industrial estates and small-scale industries. This industrial development has led to a high demand for groundwater extraction, resulting in overexploitation of groundwater and seawater intrusion. Additionally, the current metro area population of Mangalore in 2023 is 749,000, reflecting a 1.77% increase from 2022 (Mangalore, India Metro Area Population 1950–2023). The research region is located in India's Western Ghats, with the Arabian Sea on the other side. Netravati, Gurupura,

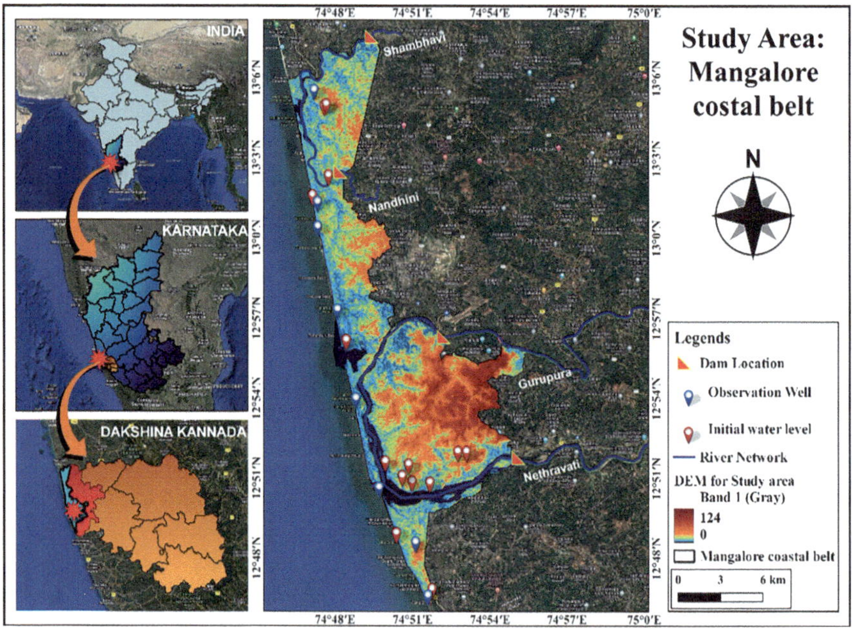

Fig. 7.1 Location of the study area

Nandhini, and Shambhavi are the four well-known significant rivers in Mangalore. They pass across the town of Mangalore and discharge into the Arabian Sea to the west (Suryawanshi et al. 2023).

7.3 Data Inventory and Methodology

Monitoring groundwater levels over time provides critical insights into flow patterns and regions affected by seawater intrusion. Data from the Central Ground Water Board (CGWB) is crucial for managing India's groundwater resources, offering key information on fluctuations that support urban planning, agriculture, and water scarcity strategies. Understanding aquifer characteristics, such as permeability, porosity, and hydraulic conductivity, is essential for predicting groundwater movement and seawater intrusion pathways. Additionally, data on pumping and recharge rates is vital for maintaining the balance between freshwater and seawater in the aquifer, as excessive pumping or insufficient recharge accelerates intrusion. Finally, mapping groundwater flow direction and velocity helps identify the pathways and rates of seawater intrusion. Table 7.1 illustrates the major dataset used in this chapter.

Table 7.1 Data procurement for the study

Data required	Sources
Field data (observation wells, water samples)	Field survey
Hydrogeological parameter (conductivity, storativity, initial head)	Field survey
DEM	USGS
Drainage pattern, vented dam	Digitized using OGIS
Surface creation	VMOD flex
LULC analysis	Landsat-8

7.3.1 Hydrogeological Data

This data is crucial for understanding the dynamics of saltwater intrusion into freshwater aquifers in coastal areas. A GPS tracker was used for the assessment of observation well data. Observation wells play a pivotal role in comprehending the complex processes of seawater intrusion within coastal aquifers. For this study, approximately eight wells located along the coastal belt of Mangalore were considered. Key hydrogeological data obtained from these wells are essential for conducting effective seawater intrusion studies. These wells, strategically positioned to monitor groundwater levels and salinity, offer a wealth of insights crucial for effective management and mitigation strategies of observation well data shown in Fig. 7.2, and Fig. 7.3 represents the graphical representation of the depth to ground water level. To monitor the extent of saltwater intrusion along the coastal belt, rivers, and estuaries, these areas must be considered in the modeling approach. The Thumbe and Adayar vented dams are located within the Nethravathi River basin.

7.3.2 Water Quality Data

Water quality analysis is a critical aspect of seawater intrusion studies, which aim to comprehend the infiltration of saline water into freshwater aquifers. By assessing various parameters, these analyses help gauge the extent and implications of seawater intrusion. Salinity, measured through electrical conductivity or total dissolved solids, is a primary indicator, alongside chloride concentration, which highlights the influence of seawater. Other factors like pH, dissolved oxygen, nutrients, trace metals, and temperature offer insights into the changing chemical composition. Groundwater level monitoring, hydraulic conductivity assessment, geophysical surveys, and modeling techniques provide a comprehensive understanding of the phenomenon, guiding effective management and preservation strategies for precious freshwater resources.

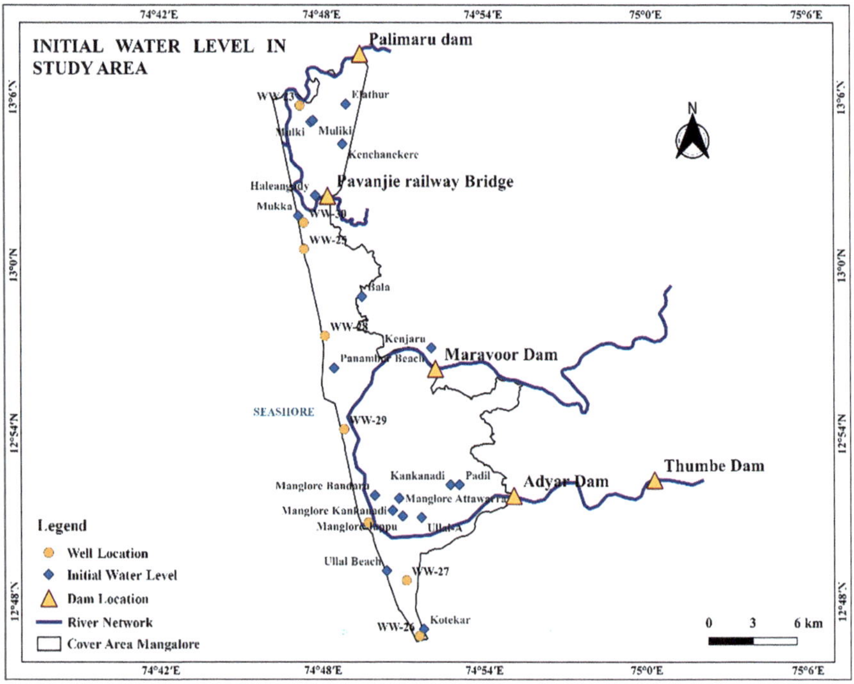

Fig. 7.2 Location of dams in the study area, including observation wells and initial water levels

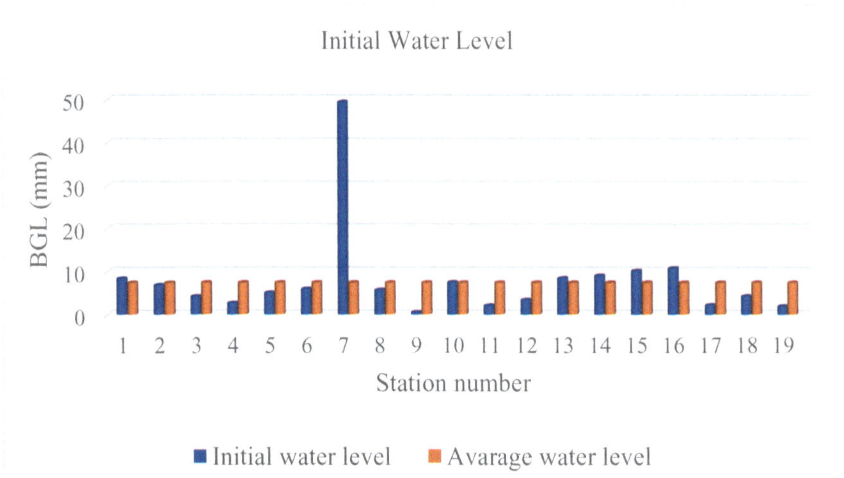

Fig. 7.3 Initial water level from CGWB monitoring wells

7.3.3 Decadal Land Use Pattern in Coastal Development

The land use and land cover in the Mangalore region have undergone dynamic changes over the last decades. Over this period, various factors such as urbanization, industrial expansion, agricultural practices, and environmental conservation efforts have influenced the transformation of the landscape (Babar and Ramesh 2015). In the year 2000, the Mangalore region predominantly featured a mix of agricultural lands, wetlands, natural vegetation, and urban settlements. Agriculture played a significant role, with extensive farmlands and plantations dotting the landscape. Coastal areas hosted fishing activities, and there were pockets of natural habitats like mangrove forests and wetlands that supported local biodiversity.

As the years progressed, urbanization became a prominent driver of change. The expansion of Mangalore's urban areas, resulting in the conversion of agricultural lands and natural spaces into residential, commercial, and industrial zones, has been analyzed in Figs. 7.4 and 7.5 for the years 2013 and 2023, respectively. The establishment of industries such as Mangalore Refinery and Petrochemical Ltd. (MRPL) and Mangalore Chemicals and Fertilizers Ltd. (MCF) contributed to the industrial footprint in the region.

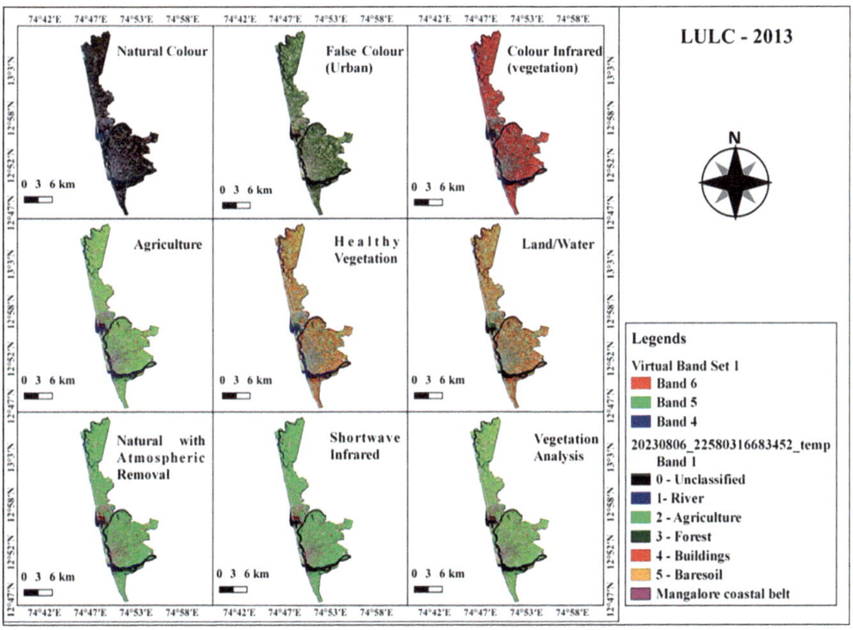

Fig. 7.4 LULC change pattern in 2013

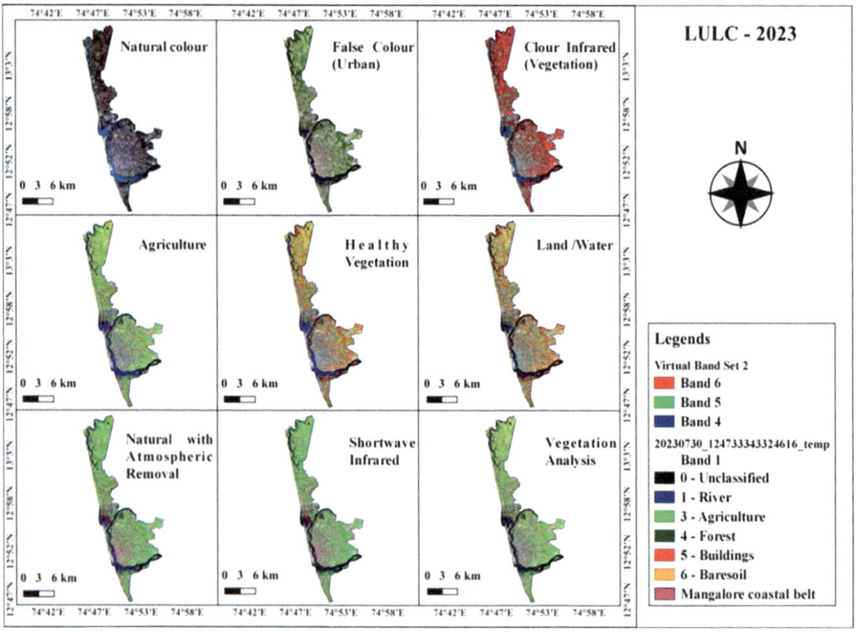

Fig. 7.5 LULC change pattern in 2023

7.3.4 Methodology to Carry Out the Study

The analysis of seawater intrusion involves a systematic and multidisciplinary approach to comprehensively understand and manage this complex phenomenon. The process (Fig. 7.6) begins with a thorough hydrogeological characterization of the study area, establishing the foundation for subsequent steps A well-designed monitoring well network is established to collect groundwater data, including salinity levels and flow patterns. These data are then utilized to develop and calibrate numerical models that simulate the movement of groundwater and the intrusion of seawater. By predicting various scenarios and assessing impacts, the analysis sheds light on potential vulnerabilities and risks associated with seawater intrusion, particularly in sensitive zones. Mitigation strategies are formulated and evaluated based on the study's findings, and engagement with stakeholders ensures that local perspectives are considered. Ultimately, the analysis provides crucial insights for sustainable groundwater management, helping to safeguard coastal aquifers and the ecosystems and communities that rely on them from the adverse effects of seawater intrusion. Analyzing and conducting experimental studies, along with preprocessing outputs, provide essential inputs for the model setup, which is considered part of the postprocessing study.

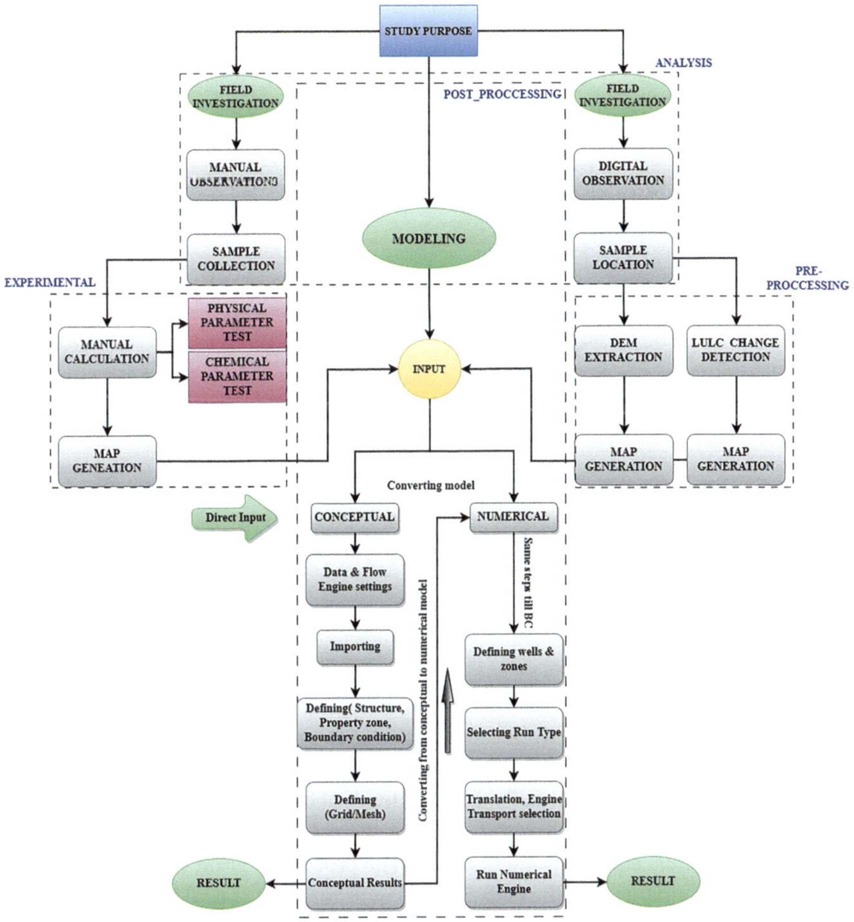

Fig. 7.6 A schematic diagram representing the study's methodology

7.3.5 Defining Model

A model is a simplified version of a complicated system or process made to understand how it works. In groundwater modeling, a model usually has a group of math equations that describe how groundwater flows or how pollutants move. These equations are solved using computers to imitate how the system acts over time. For instance, a groundwater model might have numbers for things like how easily water moves through the ground, how much empty space there is, how quickly water gets added, and the edges of the area. A sample model is represented in Fig. 7.7. This model allows for the simulation of various scenarios by adjusting these parameters to predict how changes affect groundwater flow and other subsurface hydrological processes.

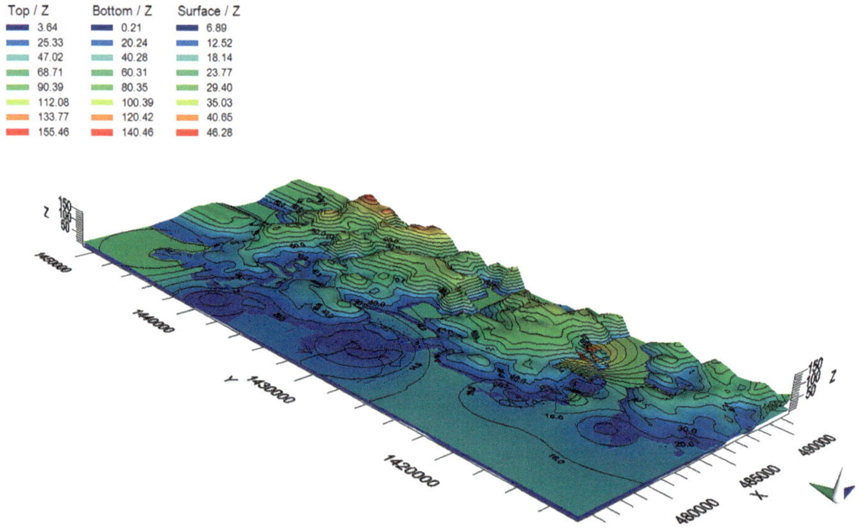

Fig. 7.7 Model domain obtained from Visual MODFLOW Flex

7.3.6 Model Simulation

The engine was configured to work seamlessly with the provided data, and by selecting a specific run type, it became practical to operate and gain insights from translation logs. Model gridding encompasses the partitioning of the model's scope into an interconnected grid of cells. Each individual cell stands as a distinct section or space within the scope. This grid assumes the role of the spatial structure upon which mathematical equations are resolved. The process of gridding entails defining the dimensions and forms of the cells, as well as arranging the grid layout. Figure 7.8 represents finite element mesh (FEM). The engine is equipped with built-in time steps, ensuring both the run and translation processes are notably swift. Following translation, the model can be set in motion for numerical modeling using the SEAWAT engine. If we have accurate information during the model construction phase and access to essential data, the project stands to thrive. The translation log serves as the final step in assembling all required packages, culminating in the initiation of the run.

7.4 Results and Discussion

The results and discussion phase of a seawater intrusion study using groundwater modeling involves presenting simulation outcomes and analyzing their implications. This step provides insights into intrusion patterns, evaluates potential impacts, and assesses the effectiveness of strategies to mitigate saltwater encroachment. By

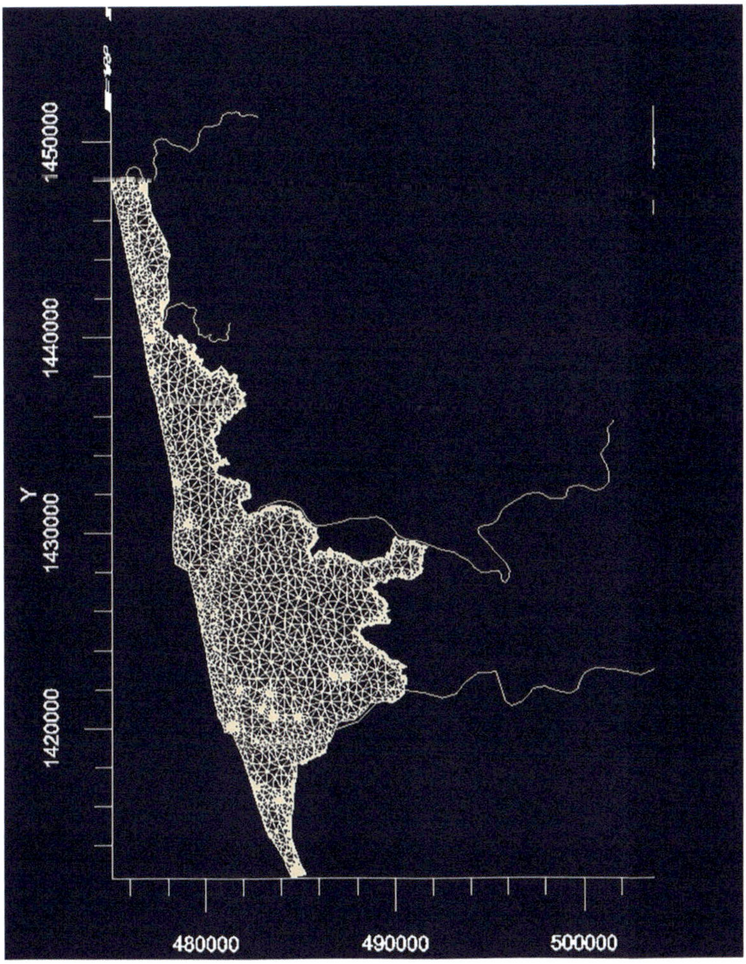

Fig. 7.8 Grid refinement (finite element mesh)

contextualizing the results, discussing their significance, and considering uncertainties, this phase informs decision-making and policy formulation and guides future research directions for ensuring the sustainable use of groundwater resources in coastal areas. Incorporating experimental results into a seawater intrusion study using groundwater modeling enhances the comprehensiveness of the analysis. By combining simulation outcomes with real-world data, a more accurate and detailed understanding of saltwater intrusion dynamics can be achieved.

7.4.1 Coastal Groundwater Flow Results

The groundwater flow results from Visual MODFLOW Flex provide a clear and detailed look at how water moves in the aquifer. They use maps and numbers to show the direction, speed, and pressure of groundwater flow. These results help us understand where water goes and how it behaves underground. We can see the pathways it follows and the areas where it moves quickly or comes together. Velocity and water table data have been combined to illustrate the flow velocity and GWT within a specific area, as shown in Fig. 7.9. The color scale represents groundwater depth: Blue/Green indicates a shallow water table, Yellow/Orange represents moderate depth, and Red/Brown indicates greater depth.

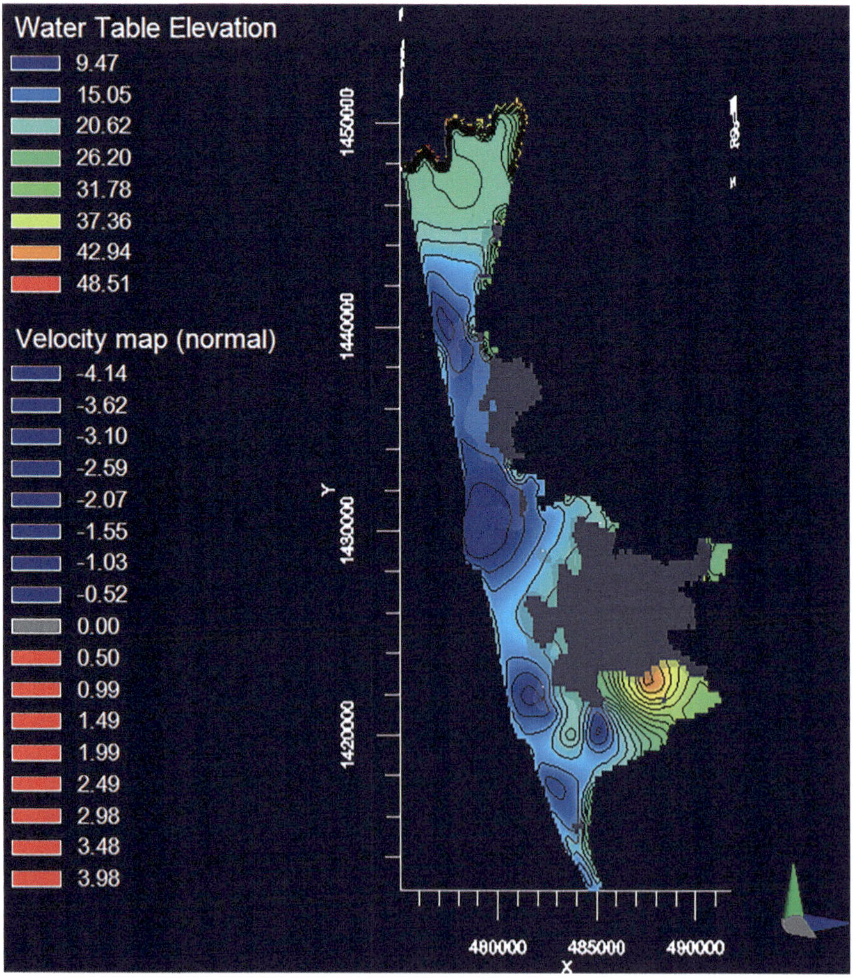

Fig. 7.9 Velocity and water table BMP map

7.4.2 Water Quality Analysis

The water quality data is presented through tables and graphs, detailing various measurements. Averages, ranges, and patterns are analyzed to interpret the data, comparing results to guidelines to assess quality and identify issues. Tables 7.2 and 7.3 list experimental measurements of physical and chemical parameters.

Groundwater's physical characteristics (Fig. 7.10), including pH, ORP, DO, EC, and TDS, offer insights into aquifer health. Water levels, varying from 5.14 to 1.25 meters, show fluctuations, with lower levels near the southern edge. pH ranges from acidic (5.57) in northern wells to neutral (7.549) in the south, indicating higher acidity in areas like Mangalore city, which may impact water quality for domestic or agricultural use. ORP measurements show oxidative conditions in the north (8.499) and reducing conditions in the south (−0.299), affecting contaminant mobility and groundwater quality. High EC and TDS in northern areas suggest seawater intrusion, indicating increased salinity due to coastal proximity and over-extraction. Lower southern water levels correlate with higher ORP and slightly acidic pH, showing complex aquifer dynamics. DO concentrations vary, with higher levels in the north possibly indicating better recharge conditions, while lower levels in the south suggest limited oxygenation due to increased organic matter or reduced aquifer replenishment. Figure 7.11 maps the spatial distribution of chemical parameters in Mangalore's coastal region.

Temperatures range from 28.36 °C to 32.55 °C, higher in southern wells. Salinity is nearly zero, indicating low saline intrusion. Magnesium levels are higher in the north, while calcium concentrations (320 to 839 mg/L) are elevated in the southern and central regions. Total hardness, ranging from 240 to 559 mg/L, is concentrated in the south. Chloride levels, between 39 and 159 mg/L, are higher in the northern and central parts. Overall water quality indexing of the Mangalore delta is illustrated in Table 7.4.

Table 7.2 Experimental results of physical parameters to check the vulnerability of collected samples from well

Well ID	GWL (m)	pH	ORP	DO (%)	DO (mg/l)	EC	TDS (mg/l)	Temp (°c)
WW-23	–	7.53	3.7	8	0.6	381	190	30.64
WW-25	2.3	6.4	−0.3	6.7	0.51	213	106	29.21
WW-26	5.15	5.86	7.8	8.9	0.61	281	141	30.37
WW-27	2.25	5.57	8.5	10.1	0.76	293	147	30.4
WW-28	2.95	7.44	3.2	9.5	0.7	491	245	32.55
WW-29	–	7.55	0.4	10.9	0.84	954	129	30
WW-30	1.25	6.77	4.5	8.5	0.67	453	226	28.36
WW-31	1.4	6.35	4	3.7	0.52	631	316	29.39

Table 7.3 Experimental results of chemical parameters to check the vulnerability of collected samples from well

Well ID	Salinity	Mg	Ca	TH	Cl	Na	K	HCo_3	SO_4
WW-23	1	−200	600	400	39.99	14.58	5.43	148	0.059
WW-25	0	−400	840	440	59.99	11.4	13.84	64	0.025
WW-26	0	−320	800	480	59.99	18.72	5.13	48	0.063
WW-27	1	0	560	560	80	55.25	9.01	32	0.072
WW-28	0	−120	360	240	99.94	29.3	5.67	136	0.046
WW-29	0	−160	560	400	159.9	61.79	8.32	208	0.122
WW-30	0	−80	520	440	119.92	28.07	7.34	96	0.095
WW-31	1	80	320	400	119.92	33.11	9.09	168	0.172

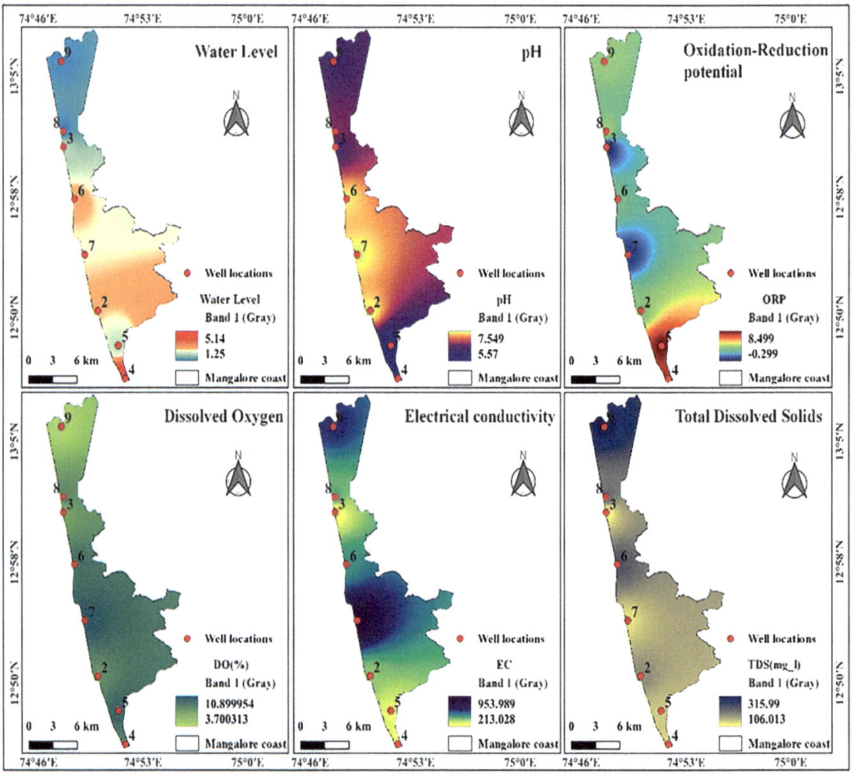

Fig. 7.10 Spatial distribution map of physical water quality parameters

7.4.3 Conceptual Model Outputs

Interpolation values extracted from software play a main role in generating contour maps. Figure 7.12 shows contours from Visual MODFLOW Flex. These maps offer a visual depiction of how a certain parameter changes across a two-dimensional surface.

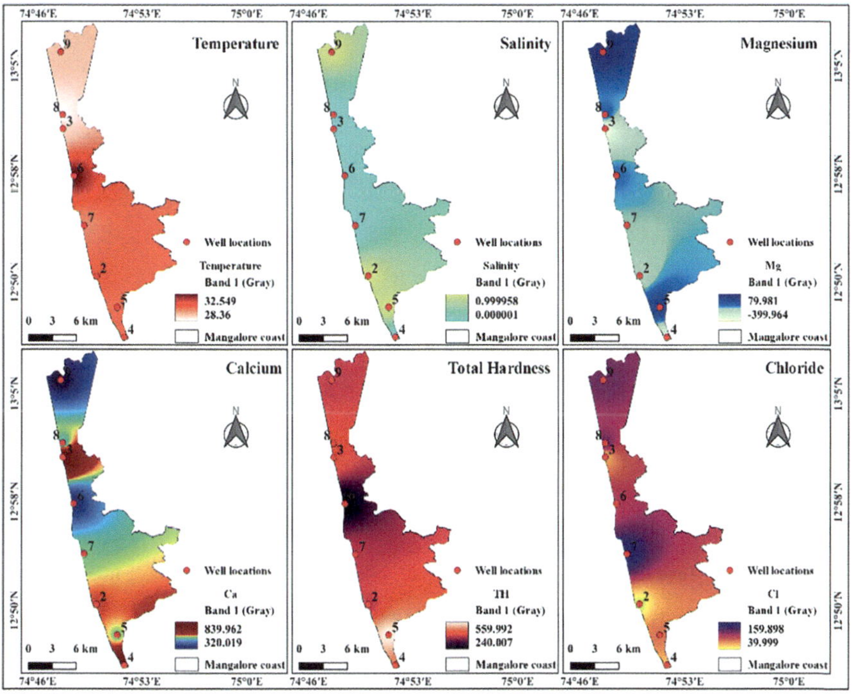

Fig. 7.11 Spatial distribution map of chemical water quality parameters

Table 7.4 Water quality analysis data obtained

Well ID	WQI	Water quality
WW-23	89.46	Good water quality
WW-25	88.31	Good water quality
WW-26	46.49	Bad water quality
WW-27	74.1	Good water quality
WW-28	80.131	Good water quality
WW-29	87.89	Good water quality
WW-30	69.83	Poor water quality
WW-31	74.507	Good water quality

These lines connect areas with the same interpolated value, allowing us to perceive the spatial distribution of the parameter. Contour maps provide an intuitive means of understanding how the parameter fluctuates within a given area. By visually representing the variations, trends, and patterns, these maps are valuable tools in fields such as geography, environmental science, and geology. They allow us to discern changes in parameters like elevation, temperature, pollution levels, and more, across landscapes. Ultimately, interpolation values and the resulting contour maps facilitate a better grasp of complex data, enabling informed decision-making and insightful analysis, as shown in Fig. 7.13.

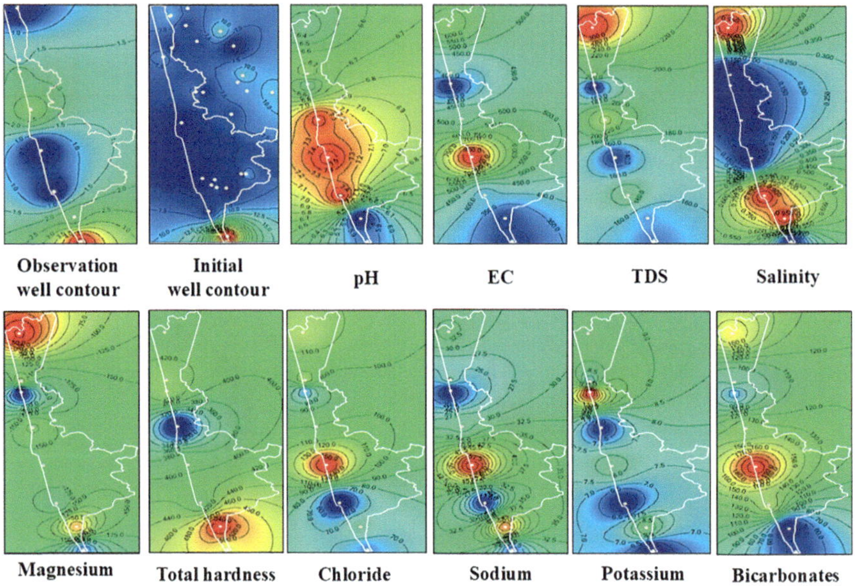

Fig. 7.12 Contour visualization from MODFLOW Flex

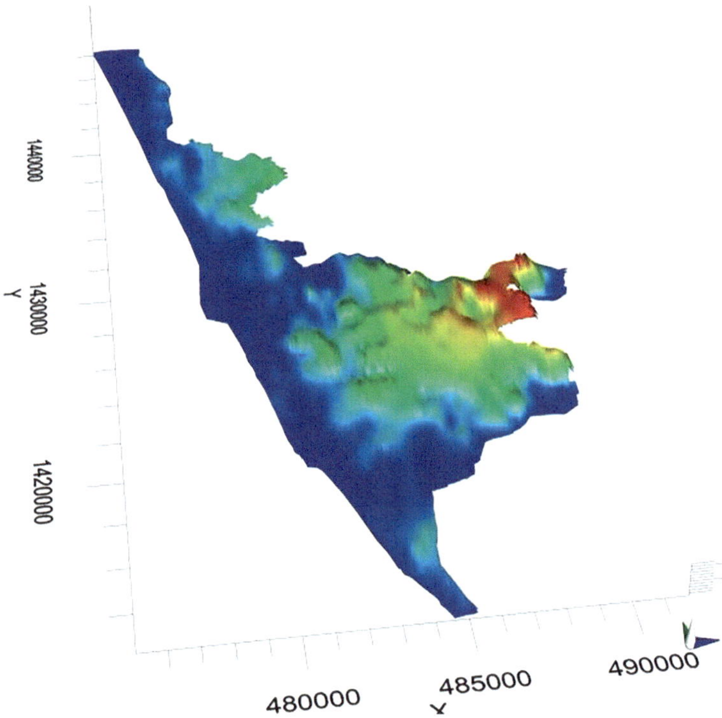

Fig. 7.13 Conceptual model domain and contours obtained using Visual MODFLOW Flex

7.4.4 Numerical Model Outputs

Based on the provided data, we were able to generate BMP files for hydraulic head, water budget, velocity, and drawdown, as illustrated in Fig. 7.14. These outputs serve as key components of the model results. With more refined and well-defined data, the study can be extended and enhanced for future applications. Velocity and water table BMP maps are available in both 2D and 3D formats. A Head BMP map

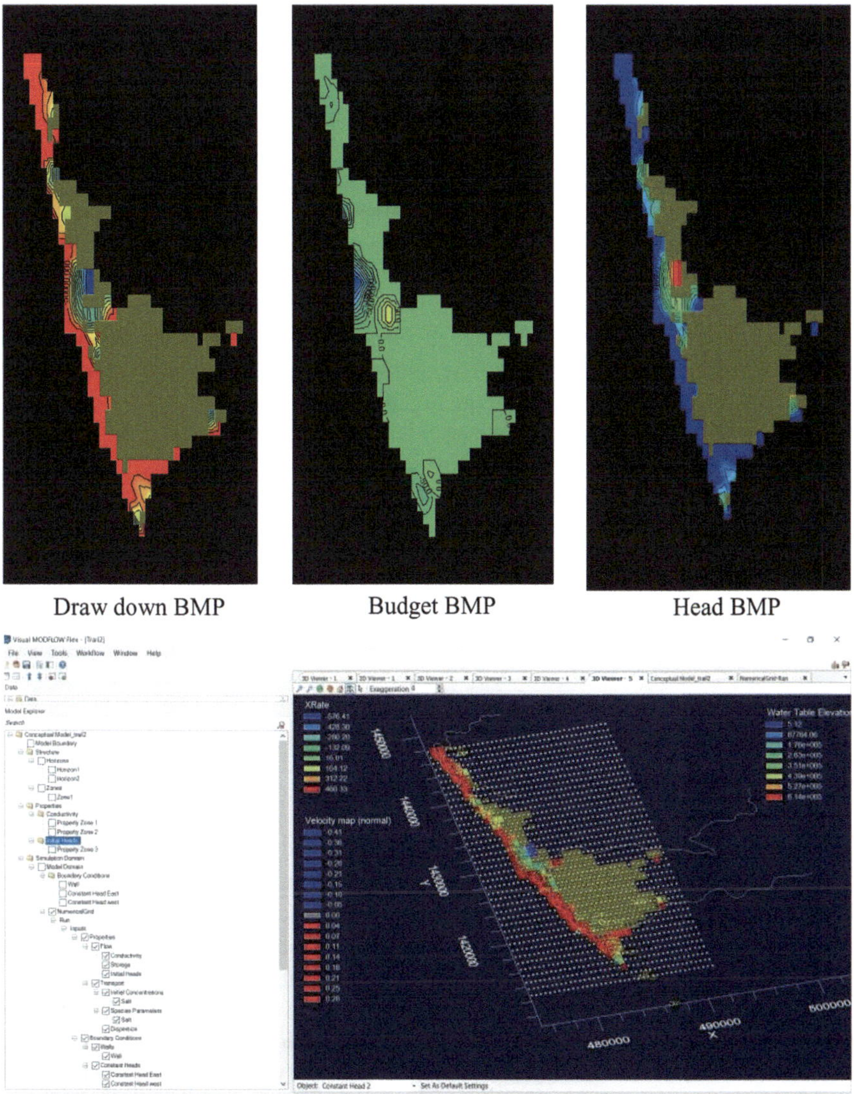

Fig. 7.14 Numerical model output as a hypothetical output of the available data

depicts water table elevation using color gradients (Blue/Green: higher elevation, Yellow/Orange: medium elevation, and Red/Brown: lower elevation). A Drawdown map displays groundwater level changes using color visualization (Blue/Green: minimal/no drawdown, Yellow/Orange: moderate drawdown, and Red/Brown: intense pumping and significant level changes). A Budget Balance Map displays key water components: precipitation, infiltration, runoff, evapotranspiration, groundwater recharge, and streamflow (blue: precipitation, runoff, and recharge; red: evapotranspiration and streamflow; green: positive balance; and yellow: negative balance and scarcity.

The seawater intrusion modeling study yielded significant findings that offer valuable insights into aquifer behavior and saltwater–freshwater interactions. The research identifies areas where saltwater intrusion is rapidly advancing or remaining static, evaluates how pumping rates and aquifer properties affect groundwater availability, and explores salinity distribution to understand mixing processes between salt and fresh water. Through sensitivity analyses, critical parameters are determined, while comparisons with field data confirm the model's precision, enhancing its reliability for sustainable groundwater management. A vulnerability assessment of well water quality uncovered problematic conditions in wells 26 and 30. Well 26, located near salt wedge estuaries in Uchila, is at risk of contamination from nearby aquaculture waste, while well 30, situated close to Pavanje estuaries in Mangalore, is susceptible to saltwater intrusion. The study employed an experimental approach to evaluate the impact on well water quality in the area using a vulnerability assessment. Results showed that wells 26 and 30 were in unfavorable conditions. These wells, located in over-exploited areas, were found to be directly exposed to contaminated industrial zones. Specifically, well 26, situated near Uchila or Uchil, which has salt wedge estuaries close to Someshwara, was vulnerable to potential contamination from industrial effluents used in nearby aquaculture operations. Similarly, well 30, positioned near the estuaries of Pavanje or Nandhini in Mangalore, close to Mukka, faced threats from saltwater intrusion due to the river's outflow into the Arabian Sea.

7.5 Conclusions

This study highlights the necessity of ongoing monitoring and periodic evaluation of water quality in coastal regions. The insights gained from this Visual MODFLOW Flex investigation serve as a valuable foundation for future studies. The software was utilized to investigate groundwater dynamics in the Mangalore delta's coastal area. The model's precision was validated by the strong correlation between predicted and measured head values. The investigation revealed that the river system plays a crucial role in influencing the shallow coastal aquifer. During dry seasons, groundwater extraction prompts river water inflow, while excess groundwater replenishes the river during monsoons. This interplay demonstrates a robust link between the river and aquifer, facilitating water exchange between surface and underground sources. Although conceptual models provide simplified depictions of system behavior, numerical models offer detailed simulations, jointly providing a comprehensive understanding for informed decision-making. The study identified seawater intrusion as a significant issue, resulting in increased aquifer salinity and decreased freshwater availability. The research emphasized the critical influence of factors such as hydraulic conductivity and pumping rates on groundwater flow and saltwater intrusion. The model's reliability was strengthened through validation with field data, ensuring an accurate representation of aquifer behavior. Future research should focus on refining input parameters through enhanced data collection techniques, such as real-time monitoring of groundwater levels, salinity, and other physical–chemical parameters. Advanced geophysical surveys and remote sensing tools could be integrated to understand subsurface conditions comprehensively. Additionally, future work could explore incorporating climate change scenarios into the Visual MODFLOW Flex model to assess long-term impacts on groundwater dynamics and saltwater intrusion. A comparative analysis with other groundwater modeling tools would also provide insights into the robustness of the results. Expanding the spatial coverage of field data and model validation will help ensure greater regional applicability.

References

Abd-Elaty I, Zelenakova M (2022) Saltwater intrusion management in shallow and deep coastal aquifers for high aridity regions. J Hydrol Reg Stud 40(February):101026. https://doi.org/10.1016/j.ejrh.2022.101026

Babar S, Ramesh H (2015) Streamflow response to land use–land cover change over the Nethravathi River Basin, India. J Hydrol Eng 20(10):05015002. https://doi.org/10.1061/(asce)he.1943-5584.0001177

Berehanu B, Ayenew T, Azagegn T (2017) Challenges of groundwater flow model calibration using MODFLOW in Ethiopia: with particular emphasis to the Upper Awash River Basin. J Geosci Environ Prot 05(03):50–66. https://doi.org/10.4236/gep.2017.53005

Datta B, Vennalakanti H, Dhar A (2009) Modeling and control of saltwater intrusion in a coastal aquifer of Andhra Pradesh, India. J Hydro-Environ Res 3(3):148–159. https://doi.org/10.1016/j.jher.2009.09.002

Elango L (2006) Numerical modelling of groundwater flow in South Chennai coastal aquifer.

Gottumukkala SB, Thotakura VN, Gvr SR, Chinta DP, Park R (2024) Balancing aquaculture and estuarine ecosystems: machine learning–based water quality indices for effective management. Environ Sci Pollut Res 31:1–17. https://doi.org/10.1007/s11356-024-34134-8

Hariharan V, Uma Shankar M (2017) A review of visual modflow applications in groundwater modelling. IOP Conf Ser Mater Sci Eng 263(3):032025. https://doi.org/10.1088/1757-899X/263/3/032025

Hussain MS, Abd-Elhamid HF, Javadi AA, Sherif MM (2019) Management of seawater intrusion in coastal aquifers: a review. Water (Switzerland) 11(12):1–20. https://doi.org/10.3390/w11122467

Kaur N, Kaur S (2019) Simulation of future groundwater behaviour in sirhind canal tract of Punjab using MODFLOW.

Mao X, Prommer H, Barry DA, Langevin CD, Panteleit B, Li L (2006) Three-dimensional model for multi-component reactive transport with variable density groundwater flow. Environ Model Softw 21(5):615–628. https://doi.org/10.1016/j.envsoft.2004.11.008

Mastrocicco M, Busico G, Colombani N, Vigliotti M, Ruberti D (2019) Modelling actual and future seawater intrusion in the variconi coastal wetland (Italy) due to climate and landscape changes. Water (Switzerland) 11(7). https://doi.org/10.3390/w11071502

Oude Essink GHP (2001) Improving fresh groundwater supply – problems and solutions. Ocean Coast Manag 44(5–6):429–449. https://doi.org/10.1016/S0964-5691(01)00057-6

Oude Essink GHP, Van Baaren ES, De Louw PGB (2010) Effects of climate change on coastal groundwater systems: a modeling study in The Netherlands. Water Resour Res 46(10):1–16. https://doi.org/10.1029/2009WR008719

Roy DK, Datta B (2017) Saltwater intrusion processes in coastal aquifers—modelling and management: a review. Desalin Water Treat 78:57–89. https://doi.org/10.5004/dwt.2017.20381

Scanlon BR, Keese K, Reedy RC (2003) Variations in flow and transport in thick desert vadose zones in response to paleoclimatic forcing (0–90 kyr): Field measurements, modeling , and uncertainties. Water Resour Res 39(7). https://doi.org/10.1029/2002WR001604

Sharan A, Datta B, Lal A, Kotra KK (2024) Management of saltwater intrusion using 3D numerical modelling: a first for Pacific Island country of Vanuatu. Environ Monit Assess 196(2):1–20. https://doi.org/10.1007/s10661-023-12245-y

Suryawanshi V, Ramesh H, Nasar T (2023) Groundwater quality evaluation in the Gurupura River Basin using GIS and remote sensing approach. In: 2023 IEEE India geoscience and remote sensing symposium, pp 1–4. https://doi.org/10.1109/InGARSS59135.2023.10490433

Suryawanshi V, Ramesh H, Nasar T (2024) Integrated ecological river health assessment of netravathi basin based on physiochemical and hydrochemical analysis. In: Babu SS, Mulangi GL, Kolathayar RH (eds) Technologies for sustainable transportation infrastructures. SIIOC 2023, Lecture notes in civil engineering, vol 529. Springer, Singapore, pp 917–929

Trenberth KE, Smith L, Qian T, Dai A, Fasullo J (2007) Estimates of the global water budget and its annual cycle using observational and model data. J Hydrometeorol 8(4):758–769. https://doi.org/10.1175/JHM600.1

Zheng C, Hill MC, Cao G (2012) MT3DMS: model use, calibration, and validation. Trans ASABE 55(4):1549–1559. https://doi.org/10.13031/2013.42263

Chapter 8
Exploring Phytoremediation Efficacy in Delta Ecosystem Aquaculture Wastewater Through Scientometrics

T. Vamsi Nagaraju, G. Sri Bala, V. Chanakya Varma, and G. Shyamala

Abstract The increase in aquaculture activities around the world has had a significant impact on food utilization. Intense inland aquaculture is becoming increasingly necessary to maintain protein supplies as natural sources like lakes and reservoirs reach their limits in producing shrimp. Leading nations in this field, South Asia uses various techniques to increase shrimp harvests. However, toxic substances in aquaculture wastewater, such as pesticides, leftover feed, antibiotics, and shrimp waste, raise questions since they might hurt the environment when released. Extensive research has been conducted to create efficient aquaculture effluent treatment techniques to overcome these problems. The recirculating aquaculture system (RAS) is frequently employed in Western countries; low-cost phytoremediation techniques are less prevalent in inland aquaculture even though they are in vogue in industrial treatments. This study examines the body of research on the phytoremediation of wastewater worldwide between 2000 and 2023, using bibliometric and textual analysis to identify trends. The findings show a consistent rise in publications, particularly beyond 2020, with China and India contributing the most. Since 2000, initiatives have been made to preserve and improve the environment, with a particular emphasis on techniques such as artificial wetlands and RAS. Biological remediation techniques have received more attention lately, with phytoremediation emerging as the most successful method. In order to improve remediation capabilities and support sustainable inland aquaculture, this report summarizes the literature and offers recommendations for future research.

Keywords Inland Aquaculture · Phytoremediation · Scientometrics · Ammonia

---

T. V. Nagaraju (✉) · G. S. Bala
Department of Civil Engineering, SRKR Engineering College, Bhimavaram, Andhra Pradesh, India

V. C. Varma · G. Shyamala
Department of Civil Engineering, SR University, Warangal, India

T. V. Nagaraju, G. S. Bala (eds.), *Delta Ecosystems*, Environmental Contamination Remediation and Management, https://doi.org/10.1007/978-3-031-97920-0_8

8.1 Introduction

As a need for food arises to solve global nutritional imbalances, aquaculture is becoming more and more commonplace worldwide. Aquaculture is being developed at different production phases in several countries, improving their food security. Around 110 million tons of aquaculture products are produced annually in Asian nations (Jiang et al. 2022). The top three Asian countries in terms of inland shrimp production for the world market are China, India, and Vietnam. Several factors, including institutional frameworks, human resources, infrastructure, market demand, and location, influence the global distribution of the aquaculture industry (Naylor et al. 2023). To increase fish output, several tactics have been used, such as putting best practices into practice, using technology, and enforcing rules. Aquaculture improves the economy and solves the food supply problem, but its effects on the ecosystem cannot be ignored. This is because raising aquatic species necessitates the exploitation of land for fish farming and the inclusion of foreign substances in the cultured water (Ahmad et al. 2022). The chemically added ingredients had many purposes, mainly maintaining the fish's health (Gatlin III and Yamamoto 2022). Unfortunately, total absorption is typically not achievable, leaving the residue as a pollutant in the water. Additionally, fish excretion through metabolism will cause waste to be generated in aquaculture water (Vamsi Nagaraju and Sunil 2024).

The overuse of inland shrimp populations as feed ingredients and water pollution from aquaculture effluents have resulted in environmental problems (Mitra et al. 2024). The ecosystem of receiving water bodies and sediments is deteriorated by organic materials found in aquaculture waste, such as uneaten feed and excrement (Mitra et al. 2024). Total dissolved solids dissolved and particulate organic matter and nutrients like nitrogen and phosphorus found in aquaculture effluent can be harmful to the environment and have an adverse effect on shrimp growth (Chiquito-Contreras et al. 2022). Oxygen depletion is a major result of aerobic bacteria's increased metabolic activity due to the enormous amounts of decomposable organic waste. Vulnerable aerobic organisms in bodies of water such as rivers and lakes suffer damage from this oxygen deficit. Sulfate reduction mechanisms predominate in anaerobic conditions, which prevents aerobic nitrification and denitrification (Deng et al. 2022). The death of sediment macrofauna, which is necessary for bio-irrigation, is caused by oxygen deprivation in these processes, further decreasing aeration in receiving water bodies. Due to insufficient microbial populations, organic enrichment in river sediments has a detrimental effect on the environment, upsetting the food chain in the river basin through the activity of methanogens and sulfate reducers (Sabadini-Santos et al. 2022). Aquaculture operations must mitigate the environmental effects of their discharge of untreated effluent if they are to remain sustainable. Sustainable treatment methods are necessary since nutrient enrichment from feedstock can potentially cause eutrophication in lakes and discharge channels. Aquaculture systems employ conventional wastewater treatment techniques, such as physical, chemical, and biological processes (Hasan et al. 2023). Nitrification or denitrification of organic materials can be facilitated by biological treatments such as fluidized bed reactors, submerged biofilters, trickling filters, and rotating biological contactors (Hasan et al. 2023; Aaron et al. 2021; Priyadharshini et al. 2021).

Over the past 20 years, there has been a noticeable advancement in the advent of RAS technology, and RAS has gained popularity recently (Maharjan et al. 2022). RAS was created to raise fish in areas with insufficient biophysical conditions, a shortage of water, inadequate water quality, and unfavorable environmental factors (Maharjan et al. 2022). When more affordable options are unavailable due to illness, accessibility to land, salinity, temperature, water supply, or environmental constraints, RAS offers an alternate production method. Nonetheless, additional elements prompted the creation and application of RAS. For instance, RASs are increasingly employed in the production cycle of salmonids, especially in the juvenile phases, before being moved into outdoor grow-out systems like cages or flow-through raceways (Maharjan et al. 2022). Livestock and feedstock production with RAS may support aquaculture in cages and net pens (Maharjan et al. 2022). RAS was created as an alternative to salmon cage rearing in North America and Europe. In order to prevent adverse effects on native species and biodiversity, RAS has also been designed to cultivate exotic fish species.

Even though RAS has the potential to produce more fish, its present contribution to aquaculture production worldwide is negligible. The lack of widespread adoption of RAS combines technological, societal, and economic obstacles (Fudge et al. 2023). Even though 89% of the world's aquaculture production is centered in South Asia, adopting RAS in these countries is comparatively delayed because of substantial technological and financial barriers (Araujo et al. 2022). Furthermore, as RASs are relatively recent additions to the aquaculture industry in many nations, societal acceptance is still necessary before they can be widely used (Maharjan et al. 2022). Most aquaculture worldwide is run by small- and medium-sized farms, which usually have limited funding. Inadequate design of systems and intricate engineering methods are two major obstacles for RAS (Araujo et al. 2022). Therefore, RAS is typically conducted in more developed nations rather than in underdeveloped nations. However, because they require a lot of energy, maintenance, and capital investment, these conventional technologies are costly. Therefore, it is crucial to create sustainable technology for treating wastewater from aquaculture.

8.2 Methodology

Scientific mapping, specifically scientometric analysis, is employed to visually represent important patterns and trends in a substantial body of literature within a certain research subject (Pham et al. 2021). This analysis relies on publications in journals included in the SCOPUS bibliometric database with a keyword1 and keyword2 of phytoremediation and wastewater, respectively. The study is concerned with a time frame limited to the years 2000 to 2023. Figures 8.1 and 8.2 display the

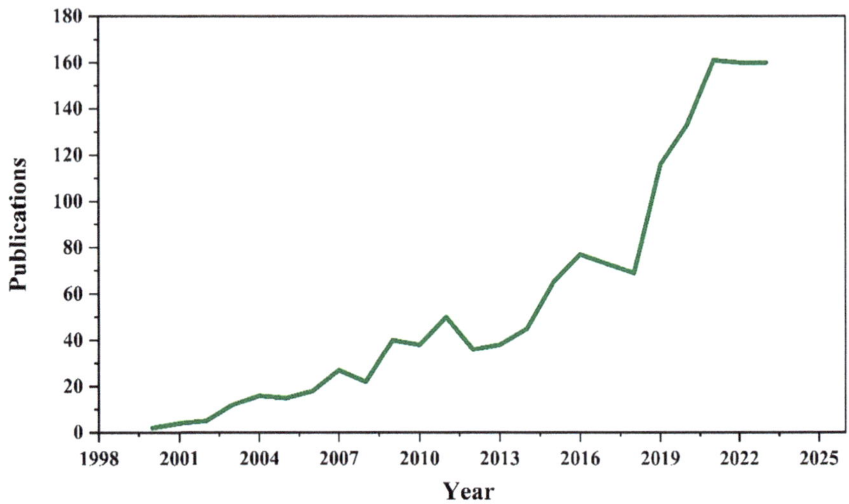

Fig. 8.1 Year-wise total publications on phytoremediation as per keyword search on the Scopus database

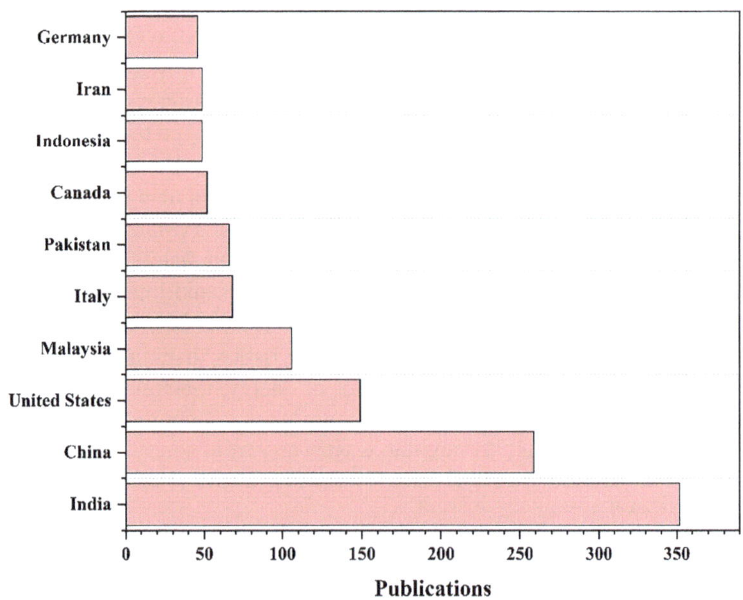

Fig. 8.2 Top-ten scientific papers contributed by countries on phytoremediation

year-on-year total publishing production from 2000 to 2023 and the contribution of top-ten countries, respectively.

Scopus was chosen in this study because of its status and it is also the most widely used scientific electronic database across the globe by numerous researchers

(Pham et al. 2021). Nevertheless, the exponential expansion of aquaculture has given rise to a multitude of issues that impede its ability to achieve long-term viability. However, the effluent treatment in aquaculture is also a burning issue that needs to be addressed since it incurs toxic heavy contaminants that affect the ecosystem badly if not discharged with proper scientific measures. Initially, microorganisms such as probiotics and bacteriophages were employed to enhance the quality of water. Furthermore, macrophytes including aquatic floating macrophytes were employed for aquaculture effluent treatment. Additionally, food crops were cultivated in aquaculture systems to purify effluent and provide marketable products. A total of 1499 documents pertaining to phytoremediation in wastewater treatment were identified between 2000 and 2023.

India and China have been the most active countries in the recent two decades in producing studies about the application of phytoremediation in wastewater water treatment and soil contamination, as seen from the Scopus database. China supplied 259 articles, while India produced 352 articles with researchers. The two Asian nations, China and India, differ significantly from the others. The interconnection of several nations indicates that efforts are being made to trace and treat wastewater and heavy metals. Based on the statistics, it was feasible to determine that China and India produce more than other industrialized countries like the United States and the United Kingdom. Research articles on the use of phytoremediation in each country can be compared to the public policies that are considered for wastewater treatment in each nation. As a result, the Indian government has been working to raise awareness of water conservation and revival through innovative programs like "Amrit Sarovar," "Nadi Utsav," and "Har Khet Ko Pani."

The scientometric analysis was used to examine the data that was obtained from Scopus, and the data were arranged and examined in accordance with the guidelines. Microsoft Excel was used to appropriately organize the recovered texts, and VOSviewer was used to create the maps (Li et al. 2021). It was feasible to determine the various topic areas, the journals that Scopus indexes, the primary keywords, the nations, and the primary authors in this subject by using VOSviewer.

8.3 Results and Discussion

8.3.1 Scientometric Analysis

A scientific article's keywords serve as a primary summarization of the findings. Since the keywords used in an article must be associated in some way, the study provides insight into the post's topic. The frequency of co-occurrence serves as an expression of this association. It is widely accepted that a word pair's frequency of occurrence in the same body of literature indicates how closely related the two subjects are. This relationship helps identify the overall direction of scientific study and the relationship between the article's contents. To identify authors and index

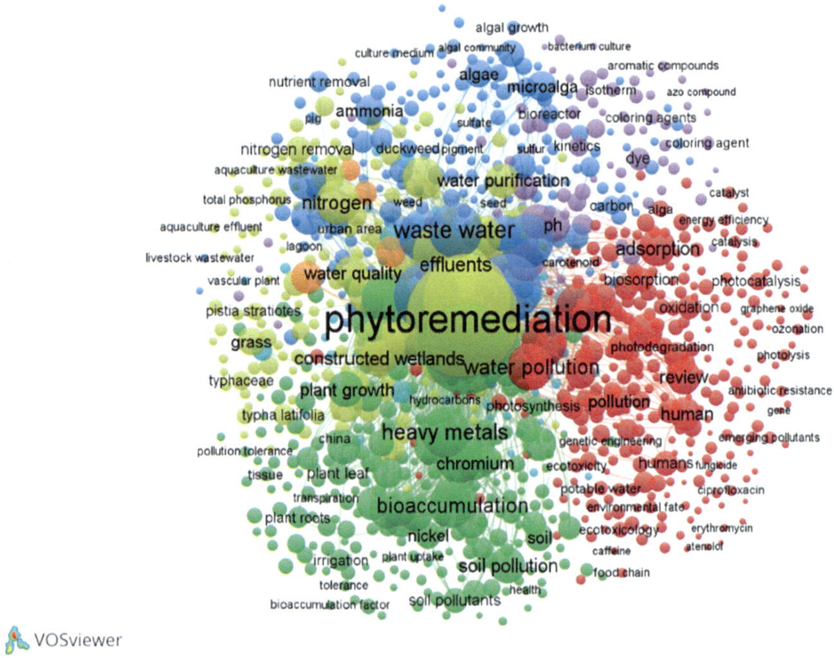

Fig. 8.3 A database of terms that frequently appear in research papers on wastewater and phytoremediation based on VOSviewer

keywords, the frequency of terms in the 1499 articles in the Scopus database was analyzed. To enhance the analysis, terms that appeared in the papers at least ten times were identified. Figure 8.3 illustrates that heavy metals, wastewater, and water contamination are the descriptors that appear most frequently in the Scopus database.

8.3.2 *Phytoremediation Techniques' Potential in Aquaculture Wastewater*

Scientometric analysis can help identify successful applications of phytoremediation in wastewater treatment, which has been widely adopted across various industries. It is particularly effective in removing ammonia. However, this type of analysis often lacks a detailed focus on aquaculture effluents, leading to incomplete data collection. Phytoremediation is a promising technology that employs living plants to decrease the toxicity of aquatic environments. The effectiveness of this approach relies significantly on the specific plants used, as they play a crucial role in the treatment process. Aquatic plants are crucial to biological wastewater treatment systems because they can be utilized for phytoremediation by different processes. Pollution eradication is dependent on several elements, including plant traits, ambient factors, time of exposure, and pollutant concentration. It is important to remember that some aquatic plant species have been successfully used in the wastewater phytoremediation process. Using phytoremediation in wetland systems can enhance the removal of contaminants, offering a straightforward and cost-effective operation. Factors like robust root systems, high biomass, and rapid growth rates are essential for the success of phytoremediation, like other wetland systems. Research on pollutant removal through phytoremediation based on plant types is summarized in Table 8.1.

Aquaculture effluents have been a significant source of phosphorous and nitrogen pollution in surface water bodies in recent years. Natural water bodies with high phosphorus and nitrogen content should be avoided since they hasten eutrophication. A high algal bloom, a decline in water quality, and an increase in the growth of aquatic weed plants are some of the negative hypoxic impacts this could have on aquatic life. Recovering and reprocessing wastewater's excess nutrients is possible to create raw materials, nutrients, and food additives used in the chemical, food processing, and agricultural sectors. The wastewater sources affect its composition, significantly affecting the process of recovering nutrients. In particular, inland aquaculture wastewater high in phosphorus and nitrogen requires appropriate treatment. Hence, inland aquaculture zones provide an essential connection between the biological and inorganic phosphorous cycles. The goal of using inexpensive biosorbent material to absorb phosphate content in the effluents is to improve treatment techniques that are consistent with the sustainable approach. Paolacci et al. (Gottumukkala et al. 2024) reported that duckweed acts as a phosphate absorption component in the effluent water treatment. Nowadays, phosphorus is a necessary component in the creation of fertilizers. Struvite, which can be used as fertilizer directly, is produced via several phosphorous wastewater recovery techniques. This method is beneficial for nitrogen-containing wastewater. As a necessary nutrient for plants, nitrogen can

Table 8.1 Phytoremediation significance in wastewater treatment

Plant type	Contaminant source	Efficiency	References
Water hyacinth	Industrial mines wastewater	95% hexavalent chromium in 15 days	Saha et al. (2017)
Macrophytes (*Pistia stratiotes* and *Hydrilla verticillate*)	Dyeing effluents	77% of chemical oxygen demand and 62% of chloride	Ahila et al. (2020)
Vetiver grass (*Chrysopogon zizanioides* (L.) Roberty)	River water	Chemical oxygen demand by 97%, total nitrogen by 90%, and phosphorus by 66%	Mahmoudpour et al. (2021)
Aquatic macrophytes (*Eichhornia*, *Pistia*, and *Typha*)	River water	90% of nutrient	Sunar et al. (2022)
Salvinia molesta	Palm oil mill effluent	95% of phosphate	Ng and Chan (2017)
Duckweed (*Lemna* sp.)	Synthetic wastewater	60% of ammonia in 2 days	Ng and Chan (2018)

be found in organic and ammonia forms in wastewater. In order to develop bioresource feedstocks useful in the manufacturing of fertilizers, phosphates, biogas, biofuels, and other valuable resource feedstock from wastewater, nitrogen recovery from wastewater is therefore viewed as a significant innovation.

After conducting a scientometric analysis on phytoremediation, it was found that there still needs to be a knowledge gap between experts or scholars from different associations and nations. Developing nations need to make the connection between intensive inland aquaculture and the ensuing effects of climate change and global warming. People must be aware that climate change is happening globally and that its effects differ from place to place and even within a single region. Thus, it is critical to advance climate change governance at all levels. Therefore, future research must determine and assess the connections between using phytoremediation in ecosystems and climate change. The application of phytoremediation to remediate mixtures of non-point source discharge pollutants involving high concentrations of nitrogenous compounds and antibiotics from intensive aquaculture has been recognized as a knowledge gap. Even though drinking storage ponds are directly connected to the water source and discharge from inland aquaculture ponds, there is still a substantial knowledge gap in phytoremediation adaption strategies, which are becoming increasingly important as local communities grow. These systems provide ecological services like stormwater utilization, wastewater filtration, and ecological water management, which have positively impacted the landscape.

8.4 Conclusions

Inland aquaculture's rise to prominence in the food production industry has made farmers more competitive. As a result of this demand, the usage of medicinal goods and inappropriate feeding techniques have put the environment at risk. In developing countries, phytoremediation treatment is being promoted as a viable substitute for RAS, which is widely used in wealthy countries.

The primary purpose of this study is to summarize the research publications conducted in this area between 2000 and 2023. There were 1499 articles found overall, which were published in 85 distinct peer-reviewed journals. Since Scopus was regarded as the primary database for wastewater treatment, it was the subject of the analysis. The researchers' primary search terms were "wastewater" and "phytoremediation." China and India have the most publications on phytoremediation because of the pollution of their water supplies, which presents many difficulties for their rapidly expanding inland aquaculture. It is crucial to emphasize that organic matter, heavy metals, and newly discovered pollutants have all been removed from industrial wastewater through phytoremediation. This work makes a substantial contribution to the corpus of current knowledge on the application of phytoremediation in wastewater treatment from inland aquaculture. It provides researchers with information about the state of the studies being conducted in this area and makes recommendations for future research projects using scientometrics.

Hybrid technology provides better therapy than independent approaches since various functionalities can be exploited in a single procedure. However, the technology is in its infancy, and more study is required to determine its efficacy and optimize the system. Therefore, several aspects need to be considered while researchers develop treatments on the pilot scale. Reducing environmental contamination is a goal that can be accomplished with the proper treatment techniques.

References

Aaron KZY, Ahmad AL, Azmi NA, Ooi BS (2021) Hybrid membrane technology for waste treatment and resource recovery from aquaculture effluent. In: Integrated and hybrid process technology for water and wastewater treatment. Elsevier, pp 565–594

Ahila KG, Ravindran B, Muthunarayanan V, Nguyen DD, Nguyen XC, Chang SW et al (2020) Phytoremediation potential of freshwater macrophytes for treating dye-containing wastewater. Sustain For 13(1):329

Ahmad AL, Chin JY, Harun MHZM, Low SC (2022) Environmental impacts and imperative technologies towards sustainable treatment of aquaculture wastewater: a review. J Water Process Eng 46:102553

Araujo GS, Silva JWAD, Cotas J, Pereira L (2022) Fish farming techniques: current situation and trends. J Mar Sci Eng 10(11):1598

Chiquito-Contreras RG, Hernandez-Adame L, Alvarado-Castillo G, Martínez-Hernández MDJ, Sánchez-Viveros G, Chiquito-Contreras CJ, Hernandez Montiel LG (2022) Aquaculture—production system and waste management for agriculture fertilization—a review. Sustainability 14(12):7257

Deng YF, Zan FX, Huang H, Wu D, Tang WT, Chen GH (2022) Coupling sulfur-based denitrification with anammox for effective and stable nitrogen removal: a review. Water Res 224:119051

Fudge M, Higgins V, Vince J, Rajaguru R (2023) Social acceptability and the development of commercial RAS aquaculture. Aquaculture 568:739295

Gatlin DM III, Yamamoto FY (2022) Nutritional supplements and fish health. In: Fish nutrition. Academic, pp 745–773

Gottumukkala SB, Thotakura VN, Gvr SR, Chinta DP, Park R (2024) Balancing aquaculture and estuarine ecosystems: machine learning–based water quality indices for effective management. Environ Sci Pollut Res 31:1–17

Hasan HA, Muhamad MH, Ji B, Nazairi NA, Jiat KW, Sim SISWA, Poh AFMS (2023) Revolutionizing wastewater treatment with microalgae: unveiling resource recovery, mechanisms, challenges, and future possibilities. Ecol Eng 197:107117

Jiang Q, Bhattarai N, Pahlow M, Xu Z (2022) Environmental sustainability and footprints of global aquaculture. Resour Conserv Recycl 180:106183

Li J, Goerlandt F, Reniers G (2021) An overview of scientometric mapping for the safety science community: methods, tools, and framework. Saf Sci 134:105093

Maharjan N, Adlin N, Hatamoto M, Yamaguchi T, Murakami Y, Araki N (2022) Advanced biological water reclamation and reuse technologies for recirculating aquaculture system. In: Clean energy and resource recovery. Elsevier, pp 51–68

Mahmoudpour M, Gholami S, Ehteshami M, Salari M (2021) Evaluation of phytoremediation potential of vetiver grass (Chrysopogon zizanioides (L.) roberty) for wastewater treatment. Adv Mater Sci Eng 2021(1):3059983

Mitra S, Khan MA, Nielsen R, Kumar G, Rahman MT (2024) Review of environmental challenges in the Bangladesh aquaculture industry. Environ Sci Pollut Res 31(6):8330–8340

Naylor R, Fang S, Fanzo J (2023) A global view of aquaculture policy. Food Policy 116:102422

Ng YS, Chan DJC (2017) Wastewater phytoremediation by Salvinia molesta. J Water Process Eng 15:107–115
Ng YS, Chan DJC (2018) Phytoremediation capabilities of Spirodela polyrhiza, Salvinia molesta and Lemna sp. in synthetic wastewater: a comparative study. Int J Phytoremediation 20(12):1179–1186
Pham HH, Dong TKT, Vuong QH, Luong DH, Nguyen TT, Dinh VH, Ho MT (2021) A bibliometric review of research on international student mobilities in Asia with Scopus dataset between 1984 and 2019. Scientometrics 126(6):5201–5224
Priyadharshini SD, Babu PS, Manikandan S, Subbaiya R, Govarthanan M, Karmegam N (2021) Phycoremediation of wastewater for pollutant removal: a green approach to environmental protection and long-term remediation. Environ Pollut 290:117989
Sabadini-Santos E, de Almeida Moreira V, de Carvalho ACB, Nascimento JR, Lopez JV, Fontana LF et al (2022) Applying enzymatic biomarkers of the in situ microbial community to assess the risk of coastal sediment. In: Microbial syntrophy-mediated eco-enterprising. Academic, pp 305–335
Saha P, Shinde O, Sarkar S (2017) Phytoremediation of industrial mines wastewater using water hyacinth. Int J Phytoremediation 19(1):87–96
Sunar S, Ali SY, Saha S, Saha P, Mukherjee P, Dutta S (2022) Nutrient removal efficiency of aquatic macrophytes in wastewater. In: Recent trends in wastewater treatment, pp 117–**143**
Vamsi Nagaraju T, Sunil BM (2024) Briefing: intensive inland aquaculture ponds: challenges and research opportunities. Environ Geotech 40(XXXX):1–5

Chapter 9
Advancing Ecological Restoration and Economic Development in Indian Coastal Areas: Strategic Approaches for Sustainability

Neetu Arya, Nimai Das, Sourabh Kumar Dubey, Pragati Singh, and Prem Shankar

Abstract India's coastal zones are essential ecosystems, rich in biodiversity and providing millions of livelihoods. Mangrove forests are a crucial component of these ecosystems, acting as natural barriers against storms and coastal erosion while also playing an important role in carbon sequestration and fisheries support. In recent years, there has been an increasing awareness of the importance of coordinated strategies that balance ecological restoration and economic development. One promising strategy is mangrove-based aquafarming, which combines aquaculture and mangrove conservation. This approach improves fish and shrimp yields through nutrient recycling, increasing coastal biodiversity, strengthening resistance to environmental stressors, and boosting the local economy. This chapter offers a comprehensive analysis of various national and international models that incorporate mangrove-based aquafarming techniques. It emphasizes the ecological and economic benefits of these practices, showing how they boost ecosystem services while supporting the livelihoods of coastal communities. The chapter also identifies the

N. Arya (✉) · P. Singh
Indian Institute of Forest Management, Bhopal, M.P., India

Forest Research Institute Deemed to be University, Dehradun, Uttarakhand, India
e-mail: naryafri21@iifmbhopal.edu.in

N. Das
Department of Environment and Developmental Economics, Indian Institute of Forest Management, Bhopal, India

S. K. Dubey
Aquatic Food Systems Specialist and Project Coordinator, WorldFish, Bhubaneswar, Odisha, India

P. Shankar
Forest Research Institute Deemed to be University, Dehradun, Uttarakhand, India

Indian Institute of Remote Sensing (ISRO), Dehradun, Uttarakhand, India

T. V. Nagaraju, G. S. Bala (eds.), *Delta Ecosystems*, Environmental Contamination Remediation and Management, https://doi.org/10.1007/978-3-031-97920-0_9

critical factors that make mangrove aquafarming a feasible and adaptable strategy across diverse environmental settings. By illustrating its ability to harmonize environmental sustainability with economic growth, the chapter highlights the significance of mangrove aquafarming in promoting resilient ecosystems and sustainable livelihoods.

Keywords Coastal regions · Biodiversity · Carbon sequestration · Water quality

9.1 Introduction

The coastal zone, where land and ocean converge, holds significant importance for human populations and the health of marine environments. Estuaries, seagrass beds, salt marshes, tidal flats, mangroves, coral reefs, and continental shelves are examples of coastal ecosystems that provide a variety of vital services. These include the supply of raw materials, food production, pollution detoxification, nutrient cycling, and habitat provision for a variety of species. Furthermore, they are key in reducing storm damage and offer spaces for recreational and leisure pursuits (Lu et al. 2018). In addition to its role as a natural barrier, the coastal ecosystem functions as an effective filtration system, preventing land-based pollutants and contaminants from entering the ocean (Ferreira et al. 2017). However, the high density of human populations in coastal areas, combined with the continuous movement of materials and energy, makes these regions particularly sensitive and vulnerable to environmental changes, both natural and human-induced. As a result, human activities, climate change, and extreme weather events have led to significant degradation of coastal environments, presenting increasing risks to these ecosystems. On a global scale, coastal marine environments are confronting numerous challenges, such as coral reef destruction, coastal eutrophication, hypoxia, wetland loss, and the accumulation of heavy metals and new pollutants from coastal exploitation (Halpern et al. 2008; Horigue et al. 2012).

Mangroves are regarded as one of the most dominating sources of the economic potential for livelihood as well as ecological assemblages of plant communities that serve as a transition zone between land and sea. Although they make up only 0.1% of the planet's surface, they control three-quarters of the world's tropical, subtropical, and intertidal coastlines. They also contribute 10–30% to the economy and store up to 10 times as much carbon per hectare as terrestrial forests (Ghosh and Mondal 2016; Aburto-Oropeza et al. 2008; Rönnbäck et al. 1999). Numerous ecosystem services are provided by mangrove ecosystems, such as provisioning (charcoal, fuelwood, and timber), regulating (controlling erosion, floods, and storms, and preventing saltwater intrusion), supporting (commercial fish species breeding, spawning, and nursery habitat), and cultural (tourism and recreation), among others. Ninety percent of mangrove forests in developing nations are in danger of going extinct as a result of overexploitation and fragmentation for extensive development operations (Bhatt and Kathiresan 2012).

Globally, mangrove forests face many risks, including timber harvesting, rising sea levels, changes in rainfall and temperature because of human interventions, and natural calamities (Alongi 2008). Climate change-related holocaust is also becoming more common in many parts of the world. Temperature is a major factor in climate change, having already risen by 1 °C, and global temperatures are expected to increase by up to 4 °C by the end of the century (Sharma and Ravindranath 2019; Canton 2021). Rising global surface temperatures increase the likelihood of more severe storms and droughts. A tropical cyclone is created when larger storms, tides, and wave currents are fuelled by the increasing amount of water vapor that evaporates into the atmosphere. The Indian subcontinent is one of the world's worst-affected regions, and the subcontinent is vulnerable to nearly 10% of the world's tropical cyclones. Each year, five to six tropical cyclones form over the Bay of Bengal and occur simultaneously on India's east coast, with two or three of them being severe. In mangrove forests, the ongoing intervention of cyclones has caused the system to decline alarmingly faster than coral reefs and upland tropical forests (Duke et al. 2007).

The decline of mangroves is a serious issue since they serve both the environment and people in many ways. According to prior research (Kathiresan and Rajendran 2005; Mafi-Gholami et al. 2020; Barbier et al. 2008), they are crucial in preventing tsunamis and storm surges along the shore. It appears that the loss of mangroves reduces protection from strong winds and flooding. The preservation of mangrove ecosystems should be given top priority due to its significance to the environment and coastal populations (Gottumukkala et al. 2024). Moreover, mangrove-based aquafarming is the best practice for the conservation and restoration of mangrove forests, so consequently we must employ it more widely. East Asian nations such as Vietnam, the Philippines, China, and Indonesia use this strategy to conserve mangroves, which helps to gain economic and environmental benefits as mangrove cover increases.

The current study centers on a mangrove-based aquafarming approach, recognized for its effectiveness, climate resilience, and eco-friendliness, promoting the sustainable growth of fish and perennial plants, particularly mangrove trees. Literature highlights that mangrove ecosystems are rapidly declining due to human exploitation and unpredictable climate shifts. This chapter aims to explore management strategies by showcasing various global and national mangrove aquaculture models, effective aquaculture practices, and their ecological and economic benefits. It also identifies key factors critical for implementing mangrove-based aquafarming in India. This approach emphasizes the reforestation of lost mangrove forests, which not only provides seasonal employment for fishermen and reduces soil erosion and marine pollution but also creates essential habitats and food sources for diverse fish and other species.

9.2 Current Status and Extent of Mangrove Degradation

India's rich coastal and marine ecosystems span a wide array of habitats, including 4992 km^2 of mangrove forests, covering 0.15% of the country's total land area. These mangrove forests are categorized based on density: 1475 km^2 (29.55%) is very dense mangrove, 1481 km^2 (29.67%) is moderately dense, and 2036 km^2 (40.78%) is open mangrove. Recent assessments by the Forest Survey of India in 2021 indicate an increase of 17 km^2 in mangrove cover, highlighting a positive trend in mangrove conservation (Ghosh and Mondal 2016). The eastern coast of India, with its expansive river deltas and alluvial deposits, supports a diverse range of mangrove ecosystems, boasting significantly greater biodiversity than the western coast. The geomorphology of each coast contributes to these differences: while the western coast is generally narrow and steep, the eastern coast is broad, allowing for more complex and extensive mangrove habitats. Mangroves on the western coast are typically found in funnel-shaped estuaries, small river mouths, and backwater areas with low tidal ranges. These mangroves are adapted to the restricted flow of water in these regions, where intertidal coastlines and sheltered bays provide suitable habitats. In contrast, the eastern coast's mangroves flourish due to the vast inflow of alluvial deposits carried by major rivers such as the Ganges, Mahanadi, Godavari, and Krishna. These deltaic regions exhibit high tidal ranges and bi-directional tidal currents that continuously bring in freshwater, supporting dense and diverse mangrove ecosystems (Bhatt and Kathiresan 2012).

Coastal ecosystems rank among the most heavily exploited and endangered natural systems across the globe (Halpern et al. 2008). The impact of human activities on these environments is severe and continues to escalate, with significant losses: 50% of salt marshes, 35% of mangroves, 30% of coral reefs, and 29% of seagrasses have either been destroyed or severely degraded worldwide (Hoegh-Guldberg and Bruno 2010). Mangroves, in particular, face significant threats across much of their range, largely due to their location in areas with growing human populations and competing coastal management priorities. Key drivers such as aquaculture, agriculture, and urban expansion have contributed to extensive mangrove deforestation, further endangering these vital ecosystems (Richards and Friess 2016). The widespread loss of mangroves is deeply alarming, not only from an ecological standpoint but also due to its detrimental effects on the economy and human welfare, especially in flood-prone countries like India. One of the key contributors to this decline is the expansion of aquaculture.

Over the past 20 years, aquaculture has grown at the fastest rate in the food industry worldwide, averaging 5.3% annually (FAO Fisheries and Aquaculture 2011). Prawn farming, a significant part of crustacean aquaculture, is primarily found along tropical and subtropical countries' coasts and provides a significant amount of foreign exchange for many developing nations in Asia and Latin America (Van Be 2000). Shrimp farming has been heavily criticized for its detrimental effects on ecosystems, biodiversity, and local communities, despite its economic advantages (Ahmed et al. 2023). Nearly half of the world's mangrove forests have disappeared since the middle of the twentieth century, with the losses being most severe in Southeast Asia, the Caribbean, and the Pacific (Truong and Do 2018).

9.3 Mangrove Integrated Aquafarming

Mangrove aquafarming refers to the practice of simultaneously cultivating some mangrove faunal species such as crabs, oysters, fishes (shrimps, pershe, vetki, tengra, baila, telapia, etc.), and floral species such as golpata on any swampy land of brackish water (Susitharan and Sindhu 2021). In order to address climate change, food security, and poverty among fishermen, mangrove aquafarming is a promising sustainable coastal agroforestry system that provides mangrove rehabilitation, fishery production enhancement, livelihood improvement, and coastal erosion prevention (Jena 2019; Hamilton 2020).

9.3.1 Process of Mangrove Aquafarming System

Mangrove seeds are usually gathered from riverbanks or discovered floating in water; July through August is the best time of year to collect them. Bamboo fencing is erected around the nursery area to provide care for the seedlings, and the seeds and seedlings need to be watered every day, generally in the morning. Before the main planting season, which starts in August and ends by the end of September, the seedlings are grown in the nursery for one to one and a half months. The typical mangrove planting density is 0.5–1 tree per square meter. Chemical fertilizers, like urea and phosphate, are frequently used to promote healthy development and guarantee a satisfactory crop. The harvested goods are taken to nearby markets to be sold after the mangroves reach maturity (Kabir and Abdul Baten 2019).

All fish and prawn species on the farm usually release their fry between December and February of each year. With very minor deviations, the harvesting season also follows the same pattern, occurring from May to June. Aquafarming takes comparatively less maintenance and care than other farming methods, particularly when it comes to feeding. The aquatic species in mangrove-based aquaculture are totally dependent on natural feed, such as mangrove leaves, and no outside food is given. It takes about 3–4 months for the majority of fish and prawn species to grow to a size that can be sold. Following harvest, the fish and prawns are dried and frozen before being delivered to nearby markets to be sold (Kabir and Abdul Baten 2019) (Fig. 9.1).

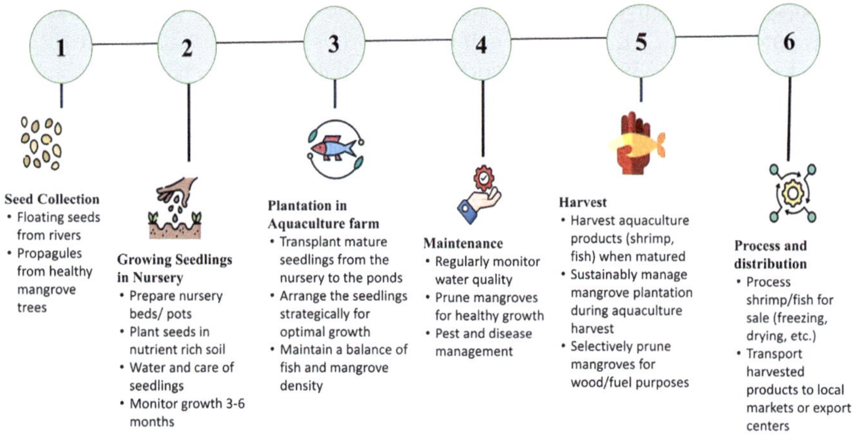

Fig. 9.1 Process of mangrove aquafarming

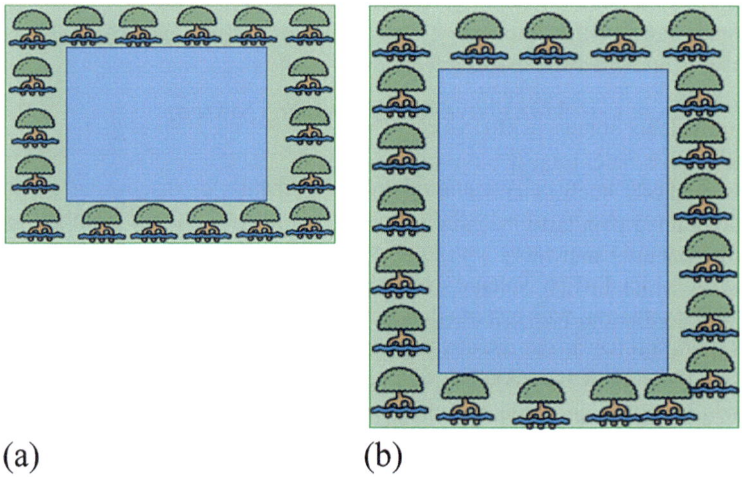

Fig. 9.2 (**a**) Peripheral model and (**b**) canal model

9.3.2 Recommended Models of Mangrove Aquafarming Worldwide

There are two aquafarming models: In the peripheral model, 70–80% of the area is covered by mangroves planted around the pond. This helps prevent unauthorized cutting, reduces soil runoff, and generates revenue from the selling of seedlings for reforestation. This model serves as the pond's organic fencing. By stabilizing the soil and fostering ecosystem health, mangroves are planted alongside canals in the canal model, providing comparable environmental and financial advantages (Susitharan and Sindhu 2021) (Fig. 9.2).

The conservation and restoration of mangrove forests have been a top priority for governments and nongovernmental organizations (NGOs) in recent years due to growing awareness of biodiversity and sea level rise. As demonstrated in Thailand in 1996 (Erftemeijer and In 2000) and the Philippines in 1982 (Bagarinao and Primavera 2005), numerous countries have previously outlawed the conversion of mangrove ecosystems. Promoting silvo-aquaculture is one strategy used by Southeast Asian nations to strike a balance between protecting mangroves and giving disadvantaged communities a means of subsistence (Sukardjo 1989; Primavera et al. 2004). Mixed mangrove–shrimp farming has been encouraged since 1978 in nations like Vietnam and Indonesia on land managed by State Forest Companies (Primavera et al. 2004). The initiative seeks to improve forest plantations, reduce erosion, and assist rural communities looking for better means of subsistence. One possible model could have been the old "gei wai" systems in Hong Kong, of which just 240 hectares are left in the Mai Po Nature Reserve (Young 2004). According to the Aquaculture Stewardship Council (ASC) guidelines, which stipulate that a shrimp farm must have at least 50% of its land covered by mangroves in order to be certified (Aquaculture Stewardship Council 2014), the province of Ca Mau in Vietnam intends to create an "organic coast" by combining mangrove–shrimp farming systems (Ha et al. 2012) (Fig. 9.3).

Mixed mangrove–shrimp production systems can be divided into three categories such as integrated, related, and separated. In the integrated system, shrimp are raised in ditches between rows of mangroves planted on platforms, with water being exchanged as needed for the management of shrimp and aquatic organisms (Van Be 2000). Both the water and mangrove regions are greater in the related system, which is sometimes referred to as "separated systems" in Vietnam (Johnston et al. 2000). Mangroves are typically located on a different platform. Shrimp ponds and mangroves, which serve as biofilters for agricultural wastewater, are separated by a dyke in the separated system (Gautier et al. 2001). There is no direct interaction between

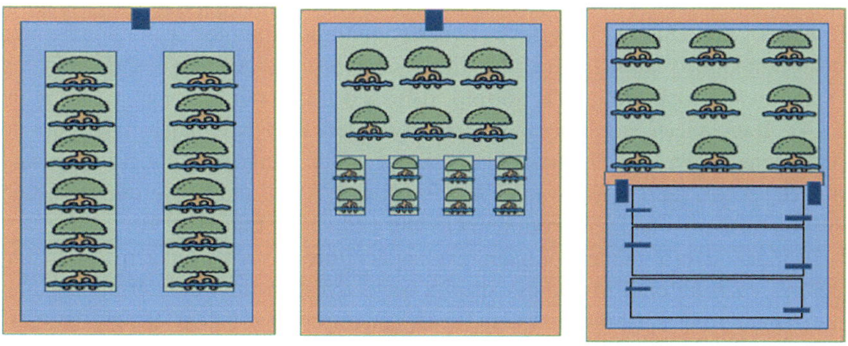

(a) (b) (c)

Fig. 9.3 (**a**) Most common integrated system, (**b**) most common associated system (in Vietnam called separated system), and (**c**) separated system applied among others in the Philippines and South America. (Bosma et al. 2016)

the mangroves and prawn farming; the pond water is controlled separately. Both the integrated and most related (komplagan) systems are categorized as type 1 silvofisheries by the Indonesian Ministry of Forestry (Van Oudenhoven et al. 2015).

9.3.3 Methods of Aquafarming Used Across the Globe and National Level

Mangrove aquafarming is one of the creative, small-scale adaptation strategies that local populations have created to safeguard their means of subsistence. Mangrove-based aquaculture techniques have become popular on a national and international scale due to their economic and environmental advantages as well as their role in mangrove conservation.

Integrated mangrove and aquaculture systems are widely practiced across various countries globally, each adopting unique models suited to their local ecological and socioeconomic conditions. In Indonesia, the Tambak System integrates mangrove forests with aquaculture to achieve multiple objectives, including food production, fuel, fodder, fertilizer, and soil stabilization. This system predominantly utilizes natural and planted mangrove species such as *Avicennia* and *Rhizophora* while stocking milkfish, tidal fish, shrimp, and natural food resources (William 2024). In the Philippines, aquasilviculture, also known as silvofishery, focuses on mangrove management, conservation, and fish production. Natural and planted mangroves are supported in these systems, which stock milkfish, mud crab, shrimp, and other tidal fish. Vietnam's aquasilviculture systems prioritize relieving land-use conflicts through mangrove rehabilitation using planted *Rhizophora*, while primarily focusing on tidal shrimp and natural food production (Aypa and Baconguis 2000).

Hong Kong employs the Gei Wai system, a type of silvo-aquaculture aimed at shrimp and fish production alongside mangrove wildlife conservation. Natural *Avicennia* mangroves are an integral part of this system (Nagaraju et al. 2022). In Thailand, community mangrove aquasilviculture promotes natural resource enhancement by planting mangroves like *Rhizophora* and *Avicennia* and managing *Nypa* forests, supporting tidal shrimp, fish, and mud crab production (On-prom 2014). Bangladesh emphasizes community-based mangrove aquasilviculture to explore innovative and climate-resilient strategies for mangrove conservation, using planted species such as *Avicennia*, *Heritiera fomes*, *Nypa*, and *Sonneratia* to sustain shrimp, mud crab, and fish production (Kabir and Abdul Baten 2019).

In India, diverse models of mangrove aquaculture are practiced across different regions, each tailored to address specific objectives. In the Sundarbans, an integrated mangrove aquaculture system focuses on improving shrimp farming, increasing the productivity of shrimp and fish, and achieving mangrove restoration and carbon sequestration. Key mangrove species include *Avicennia officinalis*, *Bruguiera gymnorhiza*, *Rhizophora apiculata*, and *Heritiera fomes*, which support the cultivation of black tiger shrimp, mud crab, and Asian seabass (Dubey et al. 2022).

In the Andaman and Nicobar Islands, aqua-silviculture systems emphasize the sustainable utilization of coastal resources without affecting native ecosystems, while generating additional employment. Mangrove species such as *Avicennia*, *Sonneratia*, *Rhizophora*, *Ceriops*, and *Cynometra* are combined with fish, mud crabs, and shrimp production (Susitharan and Sindhu 2021). Similarly, in Tamil Nadu, the Integrated Mangrove Fishery Farming System (IMFFS) aims to protect coastal areas from climate calamities and provide income generation opportunities. This system relies on mangrove species like *Avicennia officinalis*, *Bruguiera gymnorrhiza*, *Rhizophora apiculata*, and *Nypa fruticans*, supporting diverse aquaculture activities, including fish, mud crabs, and shrimp farming (Nagaraju et al. 2024).

9.3.4 Economic Empowerment and Livelihood Sustainability Through Mangrove Aquafarming

Coastal communities are often home to some of the country's most economically challenged populations. Mangrove aquafarming practices offer a sustainable and hopeful avenue for these communities by not only enhancing their economic stability but also contributing significantly to environmental conservation. These practices expand mangrove forest cover, which plays a critical role in improving fishery resources, sequestering carbon, and protecting against coastal erosion. For rural populations, especially small landholding farmers, mangrove aquafarming provides a diversified livelihood strategy. This system enables them to meet multiple needs such as timber, fuelwood, vegetables, crabs, and fish, while simultaneously supporting sustainable resource management. By integrating mangroves with aquaculture, these practices serve as both alternative and supplementary income sources, helping communities achieve financial security while maintaining ecological balance.

According to (Kabir and Abdul Baten 2019), the study explores the status, cost–benefit analysis, and environmental sustainability of CMAS (community-based mangrove aqua-silviculture) practices based on aquaculture in Bangladesh's south-western coastal area. The field survey results show that community-based mangrove aquaculture is a highly profitable technique. The average annual gross income from per bigha aquafarming yields was approximately BDT 1,83, 000, with a total expenditure of only BDT 14,750, resulting in a net advantage of approximately BDT 173250. As a result, the cost–benefit ratio is 1:12. The silvofishery technique has also been implemented in overfished and environmentally sensitive areas in Indonesia (William 2024). This technique is advantageous for silvofishery products, yielding an average net profit of $476 per acre per year. The silvofishery approach harmonizes aquaculture with mangrove preservation, ensuring that economic gains are achieved alongside ecological stability. These examples highlight the interconnectedness between livelihoods and sustainability. Mangrove aquafarming not only enhances income for vulnerable populations but also ensures the long-term health of coastal ecosystems. This symbiosis supports the broader goals of ecological restoration, climate resilience, and socioeconomic development in coastal areas, demonstrating that sustainable practices can effectively address both environmental and livelihood challenges.

9.3.5 Factors for Implementing the Mangrove Aquafarming in India

A number of ecological, social, economic, and environmental variables need to be taken into account while choosing a location for mangrove aquafarming in India in order to maximize the benefits to the environment and livelihoods and to assure sustainability. Mangrove species diversity is a crucial ecological criterion. Increased mangrove diversity improves aquaculture systems' production, habitat provision, and nutrient cycling. Another important consideration is the quality of the water; both the survival of aquatic organisms and the health of mangroves depend on ideal levels of salinity, dissolved oxygen, and pH. For nutrient availability and root growth to be supported, the sediment composition should be high in organic matter. Additionally, locations with greater potential for carbon sequestration through sediment carbon storage and mangrove vegetation have important climate mitigation benefits. Aquatic species variety and other pond biodiversity enhance ecosystem resilience and the general efficacy of aquafarming operations. From a sociological standpoint, community involvement is essential to mangrove aquafarming's long-term viability. Decision-making procedures should incorporate the communities who live close to these ecosystems, as they can offer insights into customs and knowledge that support sustainable mangrove management. By using regionally tailored mangrove ecosystem management techniques, using traditional knowledge improves site selection. Training and capacity development programs for local stakeholders also aid in improving skills, encouraging sustainable practices, and guaranteeing that the projects have a positive ecological and economic impact on the communities in which they are implemented. Economically, site selection must consider the economic benefits for locals through increased productivity of aquaculture products like shrimp and fish. The cost–benefit ratio should be favorable, ensuring that the initial investment yields long-term economic gains for the communities involved. Market accessibility is crucial to ensure that products from mangrove aquafarming can be easily transported and sold, further enhancing livelihood opportunities. Lastly, environmental and climate factors such as climate adaptation potential and resilience to natural disasters must be assessed. Sites that exhibit higher resilience to rising sea levels, storms, and other climate impacts through healthy mangrove buffers can protect both ecosystems and communities from future risks (Nagaraju et al. 2023).

9.3.6 *Benefits of Mangrove Aquafarming: Economic, Ecological, and Social*

Mangrove aquafarming is essential to the provision of several ecosystem services and activities. These include providing supplies for traditional medicine, fuel, and building materials, as well as acting as fishery habitats and feeding local communities. Mangroves are also known to help with nutrient cycling, reduce erosion, shield coastal areas from storm surges, and trap large amounts of carbon (Donato et al. 2011). Additionally, mangroves provide juvenile fish with protection from predators by serving as a nursery environment. Before moving to deeper waters, such as coral reefs, fish larvae grow in mangrove environments. Local fish communities depend on the tidal exchanges that occur within mangroves. However, these purposes are often compromised in mangrove aquafarming systems when mangroves are confined within ponds. The flow of wild fish into these places is restricted by the ponds' lack of frequent water exchanges (Nagaraju et al. 2022; On-prom 2014).

From an economic standpoint, it increases the productivity of fish and prawns by utilizing mangrove nutrient recycling, which raises yields and revenue for nearby populations. Additionally, by diversifying revenue streams through integrated forestry and fishing techniques, it supports sustainable lifestyles. Mangrove aquafarming's community-based model promotes social cohesiveness and fortifies regional governance frameworks pertaining to resource management. Additionally, it provides long-term socioeconomic security by increasing resistance to the effects of climate change, such as storm surges and rising sea levels (Nagaraju et al. 2023) (Fig. 9.4).

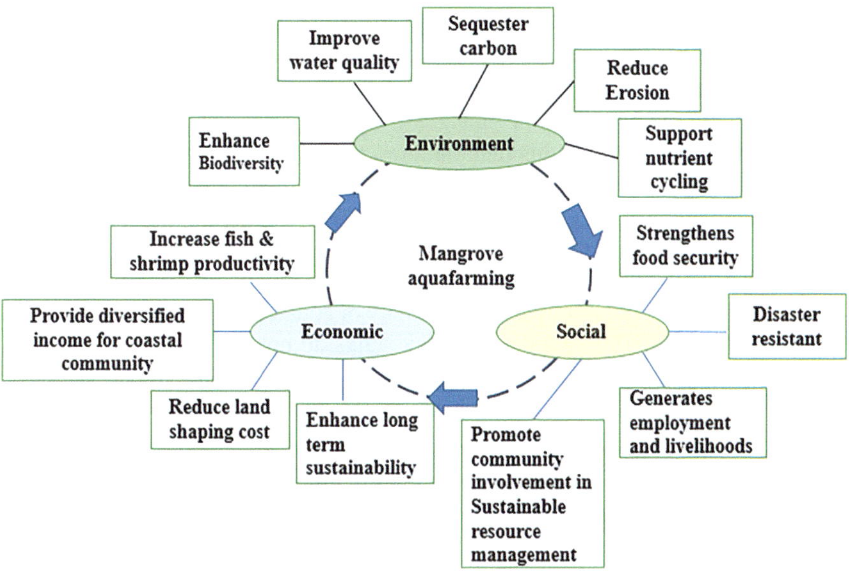

Fig. 9.4 Benefits of mangrove aquafarming

9.4 Insights for Future Considerations

To balance the growing demand for shrimp with the need for global mangrove conservation, prawn aquaculture within mangrove ecosystems should be carefully managed under controlled and eco-friendly guidelines. This approach addresses environmental concerns while promoting sustainability, ensuring that aquaculture practices align with the preservation of these critical ecosystems. Whenever possible, aquaculture activities should be relocated from fragile coastal mangrove habitats to closed inland systems, thereby reducing stress on these vital ecosystems. Rather than increasing the area of aquaculture in mangrove regions, efforts should concentrate on improving the environmental sustainability of existing ponds by adopting methods that minimize ecological damage. Mangrove aquaculture initiatives should be designed with a focus on livelihoods, while acknowledging the relatively lower biodiversity, ecosystem functions, and ecosystem services provided by these modified environments. Additionally, mangrove restoration efforts should emphasize the full recovery of interconnected and self-sustaining mangrove forests, particularly in degraded or abandoned shrimp ponds, with long-term protection measures in place. This strategy offers more ecological and biodiversity benefits compared to converting active aquaculture ponds without mangroves into mangrove-integrated systems.

9.5 Conclusions

This dual benefit makes mangrove aquafarming a valuable strategy for fostering resilience in coastal regions, especially as they face increasing environmental challenges. For India to successfully implement mangrove aquafarming in its coastal areas, several key factors must be considered. First, proper site selection is crucial to ensure the compatibility of mangrove growth with aquaculture systems. Monitoring water salinity, sedimentation rates, and tidal patterns is essential to the long-term sustainability of these ecosystems. Second, active community participation is needed, involving local fishermen and farmers in management and decision-making. Third, financial support and government policies that incentivize sustainable practices will be vital to scaling up this model. Research and development efforts should also focus on improving the productivity of aquaculture without harming the mangroves, thus ensuring that the economic gains do not come at the cost of ecological degradation. Overall, with the right policy framework and community engagement, mangrove aquafarming holds immense potential for both ecological conservation and economic development in India's coastal regions.

References

Aburto-Oropeza O, Ezcurra E, Danemann G, Valdez V, Murray J, Sala E (2008) Mangroves in the Gulf of California increase fishery yields. Proc Natl Acad Sci 105(30):10456–10459
Ahmed MU, Alam MI, Debnath S, Debrot AO, Rahman MM, Ahsan MN, Verdegem MCJ (2023) The impact of mangroves in small-holder shrimp ponds in south-west Bangladesh on productivity and economic and environmental resilience. Aquaculture 571:739464
Alongi DM (2008) Mangrove forests: resilience, protection from tsunamis, and responses to global climate change. Estuar Coast Shelf Sci 76(1):1–13
Aquaculture Stewardship Council (2014) ASC shrimp standard version 1.0. Utrecht, The Netherlands
Aypa, S. M., & Baconguis, S. R. (2000). Philippines: mangrove-friendly aquaculture
Bagarinao T, Primavera J (2005) Code of practice for sustainable use of mangrove ecosystems for aquaculture in Southeast Asia. Aquaculture Department, Southeast Asian Fisheries Development Center
Barbier EB, Koch EW, Silliman BR, Hacker SD, Wolanski E, Primavera J et al (2008) Coastal ecosystem-based management with nonlinear ecological functions and values. Science 319(5861):321–323
Bhatt JR, Kathiresan K (2012) Valuation, carbon sequestration potential and restoration of mangrove ecosystems in India. Sharing lessons on mangrove restoration 19:41–42
Bosma RH, Nguyen TH, Siahainenia AJ, Tran HT, Tran HN (2016) Shrimp-based livelihoods in mangrove silvo-aquaculture farming systems. Rev Aquac 8(1):43–60
Canton H (2021) United nations development programme—UNDP. In: The Europa directory of international organizations 2021. Routledge, pp 179–184
Donato DC, Kauffman JB, Murdiyarso D, Kurnianto S, Stidham M, Kanninen M (2011) Mangroves among the most carbon-rich forests in the tropics. Nat Geosci 4(5):293–297
Dubey SK, Censkowsky U, Saha A, Dey A (2022) A pilot of integrated mangrove-aquaculture as a nature-based solution to mitigate climate change in West Bengal, India
Duke NC, Meynecke JO, Dittmann S, Ellison AM, Anger K, Berger U et al (2007) A world without mangroves? Science 317(5834):41–42
Erftemeijer PL, In RRL (2000) Planting mangroves on intertidal mudflats: habitat restoration or habitat conversion? In: Proceedings of the ECOTONE VIII seminar "En—hancing coastal ecosystems restoration for the 213i century, Ranong, Thailand". Royal Forest Department of Thailand, Bang—Kok, Thailand, pp 156–165
FAO Fisheries and Aquaculture (2011) Aquaculture department. 2013. Global aquaculture production statistics for the year
Ferreira AM, Marques JC, Seixas S (2017) Integrating marine ecosystem conservation and ecosystems services economic valuation: implications for coastal zones governance. Ecol Indic 77:114–122
Gautier D, Amador J, Newmark F (2001) The use of mangrove wetland as a biofilter to treat shrimp pond effluents: preliminary results of an experiment on the Caribbean coast of Colombia. Aquac Res 32(10):787–799
Ghosh D, Mondal A (2016) Nature watch: in imminent jeopardy: Sundarbans in danger. Resonance 21(2):173–188
Gottumukkala SB, Thotakura VN, Gvr SR, Chinta DP, Park R (2024) Balancing aquaculture and estuarine ecosystems: machine learning–based water quality indices for effective management. Environ Sci Pollut Res 31:1–17
Ha TTT, Bush SR, Mol AP, van Dijk H (2012) Organic coasts? Regulatory challenges of certifying integrated shrimp–mangrove production systems in Vietnam. J Rural Stud 28(4):631–639
Halpern BS, Walbridge S, Selkoe KA, Kappel CV, Micheli F, d'Agrosa C et al (2008) A global map of human impact on marine ecosystems. Science 319(5865):948–952
Hamilton S (2020) Mangroves and aquaculture. A five decades remote sensing analysis of ecuadors estuarine environments, Springer Cham, 1–195. https://doi.org/10.1007/978-3-030-22240-6

Hoegh-Guldberg O, Bruno JF (2010) The impact of climate change on the world's marine ecosystems. Science 328(5985):1523–1528
Horigue V et al (2012) Marine protected area networks in The Philippines: trends and challenges for establishment and governance. Ocean Coast Manag 64:15–26
Jena JK (2019) The Asian Fisheries Society (AFS) year book-2019
Johnston D, Van Trong N, Tuan TT, Xuan TT (2000) Shrimp seed recruitment in mixed shrimp and mangrove forestry farms in Ca Mau Province, Southern Vietnam. Aquaculture 184(1–2):89–104
Kabir MH, Abdul Baten M (2019) Community mangrove aqua-silviculture (CMAS culture): an innovation and climate resilient practice by the Sundarbans mangrove forest dependent rural communities of Bangladesh. Int J Environ Clim Change 9:1–16
Kathiresan K, Rajendran N (2005) Coastal mangrove forests mitigated tsunami. Estuar Coast Shelf Sci 65(3):601–606
Lu Y, Yuan J, Lu X, Su C, Zhang Y, Wang C et al (2018) Major threats of pollution and climate change to global coastal ecosystems and enhanced management for sustainability. Environ Pollut 239:670–680
Mafi-Gholami D, Zenner EK, Jaafari A, Bui DT (2020) Spatially explicit predictions of changes in the extent of mangroves of Iran at the end of the 21st century. Estuar Coast Shelf Sci 237:106644
Nagaraju TV, Malegole SB, Chaudhary B, Ravindran G (2022) Assessment of environmental impact of aquaculture ponds in the western delta region of Andhra Pradesh. Sustain For 14(20):13035
Nagaraju TV, Sunil BM, Chaudhary B, Prasad CD, Gobinath R (2023) Prediction of ammonia contaminants in the aquaculture ponds using soft computing coupled with wavelet analysis. Environ Pollut 331:121924
Nagaraju TV, Malegole SB, Chaudhary B, Ravindran G, Chitturi P, Chinta DP (2024) Novel assessment tools for inland aquaculture in the western Godavari delta region of Andhra Pradesh. Environ Sci Pollut Res 31(25):36275–36290
On-prom S (2014) Community-based mangrove forest management in Thailand: key lesson learned for environmental risk management. In: Sustainable living with environmental risks, pp 87–96
Primavera J, Sadaba R, Lebata MJHL, Altamirano J (2004) Handbook of mangroves in the Philippines-Panay. Aquaculture Department, Southeast Asian Fisheries Development Center
Richards DR, Friess DA (2016) Rates and drivers of mangrove deforestation in Southeast Asia, 2000–2012. Proc Natl Acad Sci 113(2):344–349
Rönnbäck P, Troell M, Kautsky N, Primavera JH (1999) Distribution pattern of shrimps and fish among Avicennia and Rhizophora microhabitats in the Pagbilao mangroves, Philippines. Estuar Coast Shelf Sci 48(2):223–234
Sharma J, Ravindranath NH (2019) Applying IPCC 2014 framework for hazard-specific vulnerability assessment under climate change. Environ Res Commun 1(5):051004
Sukardjo S (1989) Tumpang sari pond as a multiple use concept to save the mangrove forest in Java. Biotrop Spec Publ 37:115–128
Susitharan V, Sindhu C (2021) Aquasilviculture: a novel safeguard for mangrove ecosystem. Justagric 1:1–8
Truong TD, Do LH (2018) Mangrove forests and aquaculture in the Mekong river delta. Land Use Policy 73:20–28
Van Be N (2000) Evaluation of coastal forest and fishery resources management strategies in Camau and Bentre Provinces in the Mekong Delta, Vietnam
Van Oudenhoven AP, Siahainenia AJ, Sualia I, Tonneijck FH, van der Ploeg S, de Groot RS et al (2015) Effects of different management regimes on mangrove ecosystem services in Java, Indonesia. Ocean Coast Manag 116:353–367
William (2024) Integrated mangrove forest and aquaculture systems in Indonesia. In: SEAFDEC/AQD institutional repository, pp 21–34. https://doi.org/9718511423
Young L (2004) Conflicts in the management of a wetland nature reserve—case study of the Mai Po Nature Reserve, Hong Kong. In: Wetlands ecosystems in Asia. Elsevier, pp 145–160

Chapter 10
Sustainable Delta Ecosystems: Modern Strategies for Achieving SDGs

G. Sri Bala and T. Vamsi Nagaraju

Abstract The sustainable future of delta regions is threatened by new global problems such as water depletion, pollutants, intense aquaculture, and climate change. A multifaceted approach to preserving and utilizing delta ecosystems is required to address these issues. In order to maintain the balance between the three dimensions and achieve a sustainable future, the holistic approach must consider not only the economic and social development but also the adverse effects of this growth on the environment. Therefore, through a critical study of their contribution to each of the Sustainable Development Goals (SDGs) and their subsets, this chapter offers a thorough viewpoint on the possible difficulties for the sustainable development of delta ecosystems. Delta ecosystems are important freshwater resources at the moment. In human cultures, they are essential for producing food, drinking water, energy, recreation, wastewater treatment, and preventing floods and droughts. Many lakes and reservoirs have suffered significant degradation due to recent intense exploitation. While eutrophication remains the primary problem for many deltas globally, habitat loss has gained significant attention recently. The sustainability of deltas can become a global issue in addition to the ongoing concerns posed by human activities like aquaculture. Regarding the SDGs, there are favorable connections between the environmental SDGs and the measures taken to achieve the sustainability of deltas. Decision-makers keep an eye on and evaluate the sustainable development of deltas based on these connections.

Keywords Delta region · Inland aquaculture · Ecosystem · Climate change · SDGs

G. S. Bala · T. V. Nagaraju (✉)
Department of Civil Engineering, SRKR Engineering College, Bhimavaram, Andhra Pradesh, India

Centre for Sustainable Inland Aquaculture, SRKR Engineering College, Bhimavaram, India

T. V. Nagaraju, G. S. Bala (eds.), *Delta Ecosystems*, Environmental Contamination Remediation and Management, https://doi.org/10.1007/978-3-031-97920-0_10

10.1 Introduction

At the meeting point of the terrestrial and aquatic environments, coastal deltas are dynamic and highly fruitful landscapes (Ibrahim et al. 2024). They are frequently densely populated because they provide chances for economic development and trade and for the formation of livelihoods, particularly through agriculture and aquaculture production (Liza et al. 2025). However, despite their high development potential, many river deltas worldwide are increasingly threatened by both natural and human-induced processes (Luo et al. 2024). Although deltas and their catchments are instinctively dynamic environments, deltaic social-ecological systems are becoming more vulnerable due to human interventions. These include inland activities such as rapid urbanization, groundwater extraction, petroleum development, intensified agriculture, and infrastructure regulation, as well as ex-situ pressures such as increased upstream water withdrawals, flow alterations, and sediment disruption. These challenges are further compounded by delta-specific hazards such as flooding, sea-level rise, climate change, and salinity intrusion (Ahmed and El-Rawy 2024). When combined, these factors worsen already-existing disparities and losses of livelihood, trigger migration waves, and frequently result in the creation of new areas of poverty, particularly in delta regions' urban centers (Sinha 2024).

Numerous networks, like the Delta Alliance1, are tackling concerns unique to populous river deltas, as the growing threats to these areas are well acknowledged (Walton et al. 2024). An "International Year of Deltas" was held in 2011 with the goal of addressing several related research problems to advance our knowledge of deltas as social-ecological structures and, consequently, support the prolonged sustainable environment of these areas (Nantavisai et al. 2024). In 2015, the International Council for Science supported a global project devoted to deltas called Sustainable Deltas 2015 (Haasnoot et al. 2024). The project highlighted the need for further research and the standardization of methodologies for studying delta environments. The recently adopted sustainable development goals (SDGs) open new avenues for learning about sustainability and development in delta areas (Zhu et al. 2024). Compared to the previous Millennium Development Goals, the SDGs place a lot more explicit emphasis on an eco-friendly environment, the effects of climate change, the value of environments and environment services, regional differences, and the need for disaggregated data. This shift presents an opportunity to collect more delta-specific data for informed decision-making and targeted action (Adem Esmail et al. 2024).

Salinization of the water and soil is another issue that affects food security and, more generally, the livelihoods of communities in many delta ecosystems (Tran et al. 2024). Yet, given the environmental sensitivity and the quickly shifting demographics of many river deltas, particularly the frequently high rates of out-migration, it is clear that more work will be needed to sustainably guarantee the socioeconomic development of delta communities (LienLe et al. 2024). Therefore, multidisciplinary research and multi-stakeholder involvement will continue to be crucial to producing solid evidence for effective policy design, even in the face of growing global political and policy development (Michaelides and Laouris 2024).

Scholars have carried out a great deal of research at many different scales, from worldwide valuations to in-depth studies of particular delta locations, due to the crucial significance of deltaic ecology facilities and the major obstacles they face (Ola et al. 2024). For instance, in the Godavari Delta, the expansion of aquaculture has been shown to impact local livelihoods by compromising ecosystem services at the regional level (Shanthakumar and Padmakumar 2024). The Godavari Delta's sediment movement is impacted by untreated aquaculture effluents and industrial contaminants, demonstrating a crucial connection between human activity and the well-being of delta regions (Aransiola et al. 2024). These investigations highlight the complex interrelationships between human activity and delta habitats, frequently highlighting a pattern in which ecological integrity is sacrificed for economic progress (Wen et al. 2024). Furthermore, the Godavari Delta demonstrates dynamic land use variations and associated financial effects on ecosystem facilities, exposing invasive species and a transition from natural saltmarshes to build regions (Ai et al. 2024). Even though research on the ecology amenities of specific deltas provides in-depth limited knowledge, these results are frequently not broadly applicable and are difficult to extrapolate to other deltas (van Doornik et al. 2024). Furthermore, this research's chronological and spatial scope is usually constrained, preventing them from accurately capturing ecological changes over extended periods of time and vast geographic areas (Diro et al. 2024). Several studies have already overcome these limitations by concentrating on research conducted worldwide (Kim et al. 2024). Human activity, land sinking, water balance, rising sea levels, and catastrophic climate conditions have increased these areas' hazards (Su et al. 2024). These growing dangers seriously threaten the vital ecosystem functions that deltas provide. These vital ecosystem services are seriously threatened by growing societal expansion and changes, such as increasing populations, upward flow disturbances, people's productivity adoption, and infrastructure growth, which affects the ecological balance and long-term sustainability of delta systems. Increasing inhabitants, urbanization, aquaculture conversion, sea level upsurge, and climate change are all putting significant socioeconomic and ecological strain on deltas, especially in South Asia, according to research on the implications of sustainable development for deltas under global change scenarios (Saleem et al. 2024). Although recent developments in the study of global deltas are encouraging, there is still a great deal to learn about the precise dynamics of land use and particular ecology facilities in the context of future events. Addressing the particular difficulties of these vulnerable locations and creating long-term, sustainable solutions require understanding these possible changes.

10.2 Sustainable Development of Delta Ecosystems

The sustainability of inland waterbodies is nevertheless ultimately threatened by anthropogenic activities, including urbanization, industrialization, and aquaculture (Numbere et al. 2024). Nearly every delta region on the planet experiences various pressures, each with intricate relationships that result in various prolonged expenses. Furthermore, the Earth's climate is anticipated to endure shifting, progressively altering its ecosystems (Saoum and Sarkar 2024). Although the effects of climate change on water resources are widespread, they can be felt locally and in diverse ways depending on the location. The consequences of climate change on delta ecosystems are mostly determined by geographic position, elevation, and morphometry (Dike et al. 2024). Small ponds in arid and tropical regions experience variations in conductivity and water level due to changes in evaporation rate and water balance. Variations in precipitation can alter the nutrients that freshwater systems need in temperate and boreal regions. It is interesting to note that two connections exist between these ponds and climate change. On the one hand, the sustainable development of deltas is impacted by the effects of climate change on temperatures worldwide, sediment movement, temporal rainfall, and the severity of hazards (Morovati et al. 2024). However, because of their decaying organic carbon store and associated releases of carbon dioxide and methane, they are both manmade and natural sources that can significantly contribute to climate change (Morovati et al. 2024).

The growing number of new toxins brought about by technological advancements that release more goods and chemicals into the environment poses another challenge to the sustainability of the delta region in the coming decades. These covert pollutants are novel substances whose effects are yet unclear and for which there is minimal environmental regulation (Bi et al. 2024). More significantly, the extensive usage of these goods in developing nations contributes to the rise in microplastics and antibiotic resistance in delta regions. Developments in agriculture, aquaculture, and livelihoods led to a high level of antibiotic resistance in ponds, which generated health issues for water consumers (Fred-Ahmadu et al. 2024).

Delta areas show a foremost role in the supply and use of freshwater, so ensuring their sustainability is essential to achieving the SDGs. Supporting a globally efficient water cycle is a primary goal in preserving water resources amid rapid population growth and the increasing impacts of climate change. Therefore, the development of deltas must promote sustainable access to sufficient amounts and attributes of various advantages for human civilization, the environment, and the economy.

10.3 SDG Interlinkages with Delta Ecosystems

The core principle of the 2030 Agenda for Sustainable Development is "leaving no one behind"—a global strategy of action for people, the environment, and prosperity. In order to eradicate extreme poverty and hunger, this Agenda's 17 Sustainable

Development Goals (SDGs), 169 goals, and 232 indicators will encourage action over the next 15 years in sectors vital to humanity and the planet (Riahi et al. 2024). Figure 10.1 highlights the 5Ps, which are included within the SDGs. The agenda created a Global Indicator Framework used in many nations to attain an eco-friendly future, even though it is not legally obligatory. However, many connections within and between the SDGs, including potential trade-offs and synergy, make it difficult for policymakers to execute them all at once. By connecting synergies and managing conflicts, countries can effectively aim and implement their policies by understanding these interlinkages (Maruf et al. 2024).

There is a pressing need for comprehensive and holistic techniques and tools to assess the nature and assets of these connections and how they impact policy and execution, as these interlinkages presently lack a strong theoretical and scientific foundation. The significance of aligning water and sanitation in the plans and strategies of other areas is highlighted by the investigation of the function these resources play in developing the three dimensions: economic, social, and environmental. From this vantage point, we closely examined a single indicator to critically examine the connections between SDG targets for the long-term sustainability of deltas globally. The interactions between the two parties are taken explicitly into account, i.e., how lakes and reservoirs contribute to the accomplishment of these 17 SDGs and how possible actions for these accomplishments affect the long-term sustainability of deltas. With scientific data backing, this study can serve as a reference for decision-makers regarding the crucial connections between the SDGs and delta ecosystem sustainability.

The connections between activities outlined in the SDG framework and their implementation are crucial for achieving a more favorable and sustainable future for lakes and reservoirs. There are two kinds of connections in question: potential conflict and synergy. Possible conflicts show the possible trade-offs and limitations of

Fig. 10.1 Delta ecosystem sustainability nexus

two acts on one another, even though synergistic interlinkages show two activities allocate a mutually helpful connection. Both direct and indirect connections are possible. Indirect interlinkages demonstrate how efforts taken to achieve sustainable development implicitly impact other areas, while direct interlinkages comprise the SDGs that specifically include ponds or perspectives directly related to lakes and reservoirs. The SDGs and lake sustainability may clash despite most interlinkages having many synergies. In particular, because they support one another, the efforts to achieve lake sustainability primarily have good ties with the SDGs about environmental factors. However, in addition to their advantages, these measures may collide with the associated social and economic SDGs. As was the case in the past, human activity's indirect effects on lake sustainability can be significant, even if most economic-driven goals seek to separate environmental degradation from economic growth. A clear example is the relationship between rising untreated wastewater levels and rising per capita income in more polluting economic activity.

10.4 Delta Ecosystem Sustainability Framework

The suggested paradigm provides crucial data to comprehend and guide the relationship between sustainability, resilience, and risk in delta areas, which may have significant ramifications for delta development in the future (see Fig. 10.2). Since these elements are interdependent, changes to one can significantly impact the others. Using the suggested methodology not only offers a thorough assessment of the distribution of climate risk for risk-based decision-making but also investigates the degree to which specific SDGs have been achieved using resilience metrics.

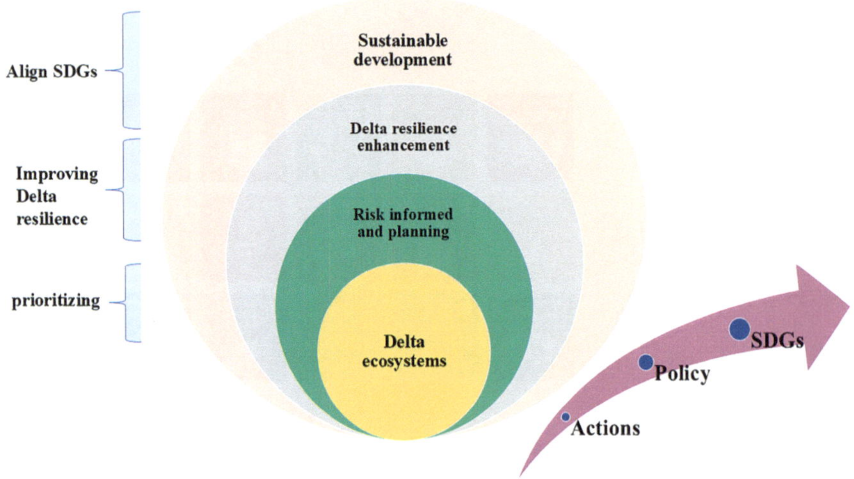

Fig. 10.2 Delta ecosystem sustainability nexus

Opportunities for multiobjective planning through risk-based decision-making may arise from this, enhancing the resilience of the higher-risk areas and ultimately facilitating sustainability. Furthermore, creating resilience is more effective than attempting to alter areas of risk outside local governments' purview, and risk assessments are fundamental to resilience management frameworks. Thus, it is beneficial to incorporate risk assessment, as done in this work, into delta ecosystem design in order to improve resilience. When deciding how to improve community resilience and advance specific SDGs concurrently with risk distribution, the resilience-based SDG attainment status results offer a comparative assessment of various locations. Coastal and delta planning and designs bear similar responsibility for mitigating disaster-causing dangers, even when climate change may affect the severity of temperatures. A better understanding of the spatial distribution of risk will help management policies and risk mitigation initiatives better safeguard communities at risk while guiding the sustainability process as a whole.

10.5 Conclusions

Since ancient times, delta ecosystems have been used for various purposes, including agriculture, aquaculture, biodiversity, climate change mitigation, and canal networks. As a result, they are essential to human health, economic expansion, and social progress. However, their future is in jeopardy because of human overexploitation and new worldwide concerns like biological invasions, climate change, land-use dynamics, and saline intrusion. In most deltas where aquaculture predominates, eutrophication has resulted in the loss of biodiversity and delta ecosystems. Additionally, each year, microbial contamination from operations involving delta ecosystems causes sickness in South Asian countries.

The threat to delta ecosystems and the need to integrate their roles in economic, social, and environmental aspects of sustainable development led to an investigation into how these ecosystems relate to the SDGs. The contributions of delta ecosystems to accomplishing these 17 SDGs were specifically emphasized, as was the effect of possible measures for these accomplishments on the sustainable development of deltas. The 5Ps and the measures taken to ensure the sustainable development of lakes and reservoirs are closely related. Due to their reciprocal reinforcement, the activities taken to achieve the sustainability of delta ecosystems mostly have good links with the environmental SDGs. These indicators concentrate on pressing issues like biodiversity, aquaculture, eutrophication, salinity intrusion, water quality, and local management. Policymakers can use these indicators to monitor the evolution of delta ecosystems and, as a result, make decisions about multifaceted, comprehensive management and policies for the deltas' sustainable future.

References

Adem Esmail B, Cortinovis C, Geneletti D, Inostroza L, Peters R, Romelli C et al (2024) Mapping and analyzing ecosystem services hotspots and coldspots for sustainable spatial planning in the Greater Asmara Area, Eritrea. Environ Manag 75:51–567

Ahmed AKA, El-Rawy M (2024) The impact of aquifer recharge on groundwater quality. In: Managed aquifer recharge in MENA countries: developments, applications, challenges, strategies, and sustainability. Springer, Cham, pp 207–222

Ai J, Chen L, He H, Han X (2024) Revealing the long-term impacts of plant invasion and reclamation on native saltmarsh vegetation in the Yangtze River estuary using multi-source time series remote sensing data. Ecol Eng 208:107362

Aransiola SA, Zobeashia SLT, Ikhumetse AA, Musa OI, Abioye OP, Ijah UJJ, Maddela NR (2024) Niger delta mangrove ecosystem: biodiversity, past and present pollution, threat and mitigation. Reg Stud Mar Sci 75:103568

Bi Z, Wang W, Zhao L, Wang X, Xing D, Zhou Y et al (2024) The generation and transformation mechanisms of reactive oxygen species in the environment and their implications for pollution control processes: a review. Environ Res 260:119592

Dike EC, Amaechi CV, Beddu SB, Weje II, Ameme BG, Efeovbokhan O, Oyetunji AK (2024) Coastal vulnerability index sensitivity to shoreline position and coastal elevation parameters in The Niger Delta region, Nigeria. Sci Total Environ 919:170830

Diro A, Kaisar S, Vasilakos AV, Anwar A, Nasirian A, Olani G (2024) Anomaly detection for space information networks: a survey of challenges, techniques, and future directions. Comput Secur 139:103705

Fred-Ahmadu OH, Ahmadu FO, Adedapo AE, Oghenovo I, Ogunmodede OT, Benson NU (2024) Microplastics and chemical contamination in aquaculture ecosystems: the role of climate change and implications for food safety—a review. Environ Sci Eur 36(1):181

Haasnoot M, Di Fant V, Kwakkel J, Lawrence J (2024) Lessons from a decade of adaptive pathways studies for climate adaptation. Glob Environ Chang 88:102907

Ibrahim MI, Bassiouni E, El-Ghareeb R, Shehata A, El Atfy H (2024) Environmental and vegetation dynamics through the Oligocene to Early Miocene of North Africa (Egypt). Geobios 82:31–51

Kim K, Lee K, Kwon ON (2024) A systematic literature review of the empirical studies on STEAM education in Korea: 2011–2019. In: Disciplinary and interdisciplinary education in STEM, pp 103–117

LienLe TH, Dang KK, Moss J, Nguyen HQ, Kristiansen P (2024) Agricultural transformation in the Vietnamese Mekong Delta in the new era. In: The Mekong River Basin. Elsevier, pp 559–617

Liza JI, Majumder SC, Rahman MH (2025) Scrutinizing the impact of blue economy factors on the economic growth in Bangladesh: an empirical study. Mar Policy 173:106542

Luo J, Straffelini E, Bozzolan M, Zheng Z, Tarolli P (2024) Saltwater intrusion in the Po River Delta (Italy) during drought conditions: analyzing its Spatio-temporal evolution and potential impact on agriculture. Int Soil Water Conserv Res 12(3):714–725

Maruf M, Chang YC, Yang L (2024) Managing institutional interlinkages for the protection of marine environment in the East Asian seas region and beyond: the case of marine plastic pollution. Ocean Coast Manag 255:107232

Michaelides M, Laouris Y (2024) A cascading model of stakeholder engagement for large-scale regional development using structured dialogical design. Eur J Oper Res 315(1):307–323

Morovati K, Tian F, Pokhrel Y, Someth P, Shi L, Zhang K et al (2024) Fishery and agriculture amidst human activities and climate change in the Mekong River: a review of gaps in data and effective approaches towards sustainable development. J Hydrol 644:132043

Nantavisai M, Hashimoto S, Meraj G (2024) Unraveling changes in deltas: exploring drivers and consequences through the lens of the IPBES conceptual framework. Sci Total Environ 175865:175865

Numbere AO, Kika EP, Izah SC (2024) Challenges and mitigation of water resource management in rural areas in Nigeria. In: Water crises and sustainable management in the global south. Springer, Singapore, pp 373–390

Ola I, Drebenstedt C, Burgess RM, Mensah M, Hoth N, Külls C (2024) Remediating oil contamination in The Niger Delta Region of Nigeria: technical options and monitoring strategies. Extr Ind Soc 17:101405

Riahi MR, Sadeghi SH, Mirnia SK, Sadoddin A (2024) Coincidence of sustainable development indicators for the nekarood watershed with the United Nation's sustainable development goals. Sci Total Environ 917:170177

Saleem A, Anwar S, Nawaz T, Fahad S, Saud S, Ur Rahman T et al (2024) Securing a sustainable future: the climate change threat to agriculture, food security, and sustainable development goals. J Umm Al-Qura Univ Appl Sci 10:1–17

Saoum MR, Sarkar SK (2024) Monitoring mangrove forest change and its impacts on the environment. Ecol Indic 159:111666

Shanthakumar M, Padmakumar V (2024) Mangroves: a mitigator of natural catastrophes. In: Mangroves in a changing world: adaptation and resilience. Cham, Springer, pp 27–44

Sinha BRK (2024) Introduction: a broad perspective on the concepts of urban dynamics, environment, and health. In: Dynamics U (ed) Environment and health: an international perspective. Springer, Singapore, pp 3–79

Su WR, Chen YH, Fu HS, Chang TY, Chen WB (2024) Assessing the inundation risk of cultural heritages along the southwestern coast of Taiwan: present and future. Reg Environ Chang 24(2):56

Tran T, Duong DV, Loc HH, Đinh VN, Trung TA, Ho VTT et al (2024) Integrated assessment of salinity intrusion and water scarcity in Ben Tre, Mekong Delta: implications for sustainable livelihoods and adaptive water management. Environment, development and sustainability, pp 1–30

van Doornik TJ, Jung MP, van Loon-Steensma JM (2024) Finding common ground: a comparison between coastal nature-based solutions in The Netherlands and British Columbia, Canada. Nature-Based Solutions 6:100173

Walton RE, Salgado J, Bisht T, Large AR, Henderson AC, Renaud FG et al (2024) Building resilience in south and southeast Asian mega-deltas. In: Resilience and riverine landscapes. Elsevier, pp 425–444

Wen L, Li B, Peng Y, Zhou Y, Weng A, Jin Y et al (2024) Exploring the evolution of coupled natural-cultural ecosystem services and their geographically scaled driven modeling in a coastal city of Southeast China. J Environ Manag 361:121265

Zhu Q, Huang SZ, Koompai S (2024) Digital transformation as a catalyst for green innovation: an examination of high-tech enterprises in China's Yangtze River Delta. Sustainable Futures 8:100277

Chapter 11
Building Sustainable Delta Futures: Policy Pathways for Conservation

G. Sri Bala and T. Vamsi Nagaraju

Abstract Most deltas face greater exposure, vulnerability, and risks associated with natural hazards due to growing population pressure, intense aquaculture, urbanization, and the effects of global climate change. Developing policies and plans for preparedness, mitigation, and adaptation requires mapping deltas' risk profiles and sensitivity. A socioecological systems viewpoint is not usually systematically incorporated into vulnerability and risk assessments, which might result in incomplete evaluations because they primarily focus on social aspects. Within risk assessment frameworks, ecosystem services, which connect ecosystem functioning and human well-being, can be utilized to describe the interdependencies between the environment and society more accurately. Therefore, we suggest an updated indicator-based paradigm for socioecological ecosystems in delta habitats, building on the vulnerability and risk evaluation systems that are currently in use. This framework is backed by a list of ecosystem service indices that were found through a thorough literature review. This updated framework helps address risk and vulnerability in coastal deltas. It makes it possible to evaluate various hazards and risks to socioecological systems both inside and outside deltas and to develop management strategies and policies that are more specifically focused on lowering the risks associated with natural hazards.

Keywords Delta region · Inland aquaculture · Delta conservation · Policy framework

G. S. Bala · T. V. Nagaraju (✉)
Department of Civil Engineering, SRKR Engineering College, Bhimavaram, Andhra Pradesh, India

Centre for Sustainable Inland Aquaculture, SRKR Engineering College, Bhimavaram, India

T. V. Nagaraju, G. S. Bala (eds.), *Delta Ecosystems*, Environmental Contamination Remediation and Management, https://doi.org/10.1007/978-3-031-97920-0_11

11.1 Introduction

Considering the rapid pace at which the socioeconomy is developing, sustainable development might help people actively look for answers to problems in social, economic, and environmental problems, thereby tackling new global challenges as a group (Salmivaara et al. 2016). Issues like resource overexploitation, pollution, climate change, and biodiversity loss have surfaced worldwide due to human activity's continued intensification. These issues have constantly strained aquaculture and land, progressively increasing the detrimental effects on ecosystems' resilience and sustainability (Nagaraju et al. 2022). In recent years, research on ecosystem services has emerged as a frontier and priority among global geography, ecology, and allied fields, serving as a link and bridge between natural processes and human activity (Wang et al. 2021). To allocate resources and coordinate benefits, ecosystem services fall into four main categories such as provisioning, regulating, social, and supporting (Syrbe and Walz 2012). These services connect ecosystems to human well-being and are crucial instruments for optimizing land use and directing planning choices based on the idea of ecological civilization. Cultural ecosystem services have been problematic in their development since they are subjective, intangible, and non-economic (Chan et al. 2012). However, they represent the deep connections between human spatial views and ways of life reproduced in outdoor restoration, scientific research, creative expression, and other industrial and consumer activities. They are the fundamental community and cultural assistance that ecosystems offer to humanity (Wu 2013). In order to achieve a rational use of resources, cultural inheritance, and innovation, protect the long-term sustainability of ecosystems, and create a peaceful existence between humans and nature, it is crucial to protect and manage cultural ecosystem services effectively (Wood et al. 2018).

One of Earth's most significant functional systems and geographical features is the delta region. Ecologically speaking, coastal regions are home to some of the planet's most biodiverse ecosystems, including significant ecosystems like agriculture, canal networks, and delta wetlands, which serve as habitats and breeding grounds for numerous species and are essential to preserving the environmental equilibrium and biodiversity of the world (Oro and Martínez-Abraín 2023). Economically speaking, delta regions are frequently vital hubs for port transit, international trade, and fishing operations. They also offer a wealth of fishing resources and serve as crucial grounds for a number of different businesses, including shipping and tourism. Socially speaking, delta regions are often highly inhabited, with many cities and settlements with rich cultural and historical traditions (Kuenzer and Renaud 2012). Compared to upland regions, delta areas' social activities, cultural legacy, and community life frequently exhibit unique traits. However, several issues also affect delta regions, such as extensive aquaculture, natural disasters, environmental contamination, and the rise in sea levels due to global change (Sekovski et al. 2012). These issues threaten global development and the environment and impact sustainable development locally. In order to preserve delta cultural ecosystem services from the standpoint of public governance, preserve their distinctive ecosystems and community values, encourage public opinions toward environmental sociability, and ultimately improve understanding of delta ecosystems, it is necessary to pay more consideration to the ecological, financial, and social growth of delta regions (Bai et al. 2016). These initiatives are essential to human health and welfare.

Numerous academic fields, including geographical area, ecology, the economy, and psychology, are involved in research on cultural ecosystem services (Pröbstl-Haider 2015). It stresses the integration of human society with nature in addition to natural elements, including biodiversity, quality of soil, and water resources (Mondal and Palit 2022). Cultural ecosystem services research, based on the "perception-attitude, values-worldview" framework, aims to comprehend the behavioral preferences and perceptual variations of various actors in their interactions with natural ecosystems to support and guide land space optimization theoretically (Bousquet and Le Page 2004). Cultural services are now given relatively little consideration in research on ecosystem services in delta regions, which primarily concentrate on providing, controlling, and sustaining services (Moberg and Rönnbäck 2003). This thorough study encourages the balanced growth of society in terms of community, cultural, and ecological facets and aids in creating more all-encompassing sustainable development methods. Scientific provision for the sustainable growth of social culture can be provided by providing a scientific source for protecting the environment, social tradition, community stability, and other aspects by thoroughly comprehending the role of cultural ecosystem services in sustainable development. In this regard, studies on delta cultural ecosystem services offer significant policy advice for decision-makers in addition to assisting scientists in investigating the connection between ecosystems and culture (Cheng et al. 2019).

The creation and use of cultural ecosystem services have received more attention lately. Rich natural resources, a delicate biological environment, and high human activity define delta areas as a unique zone of estuary interaction (Brondizio et al. 2016). In addition to reflecting the inherent ecological features of certain geographic locations, such as coasts, wetlands, and coral reefs, delta cultural ecosystem services are also strongly tied to the local populace's customs, economic activity, and way of life. The extent and long-term viability of cultural ecosystem services are also greatly impacted by variables like the degree of economic growth, rapid population growth, urban planning, and policy direction in delta areas (Hossain et al. 2016). Assessing delta cultural ecosystem services on distinctive delta ecosystems and coastal communities is crucial and essential. Nevertheless, there is a dearth of thorough investigation and systematic examination of delta cultural environmental services, and most recent research focuses on quantifying assessments of these services (Nicholls et al. 2016). As a result, the chapter's focus on the delta cultural ecosystem services' trend is poorly understood. There is a pressing need for more thorough research on delta cultural ecosystem services to uncover their diversity, complexity, and potential value to exploit better and broaden their application possibilities. Thus, this study employs knowledge mapping tools, like VOSviewer, to measure and understand the analysis, systematically evaluate and condense the delta ecosystem services research, and present future research in the Scopus database. In order to support the beneficial linkages between natural and social ecosystems, this work attempts to offer practical investigations for delta cultural ecosystem services and to develop policy frameworks.

11.2 Interactive Relationships among Delta Ecosystems, Human Activities, and Natural Resources

Delta regions are closely intertwined with human well-being, natural resource management, and cultural ecosystem services. Delta ecosystems provide rich ecological services, including biodiversity preservation, scenic beauty, and cultural heritage, and they have a direct influence on human spiritual, societal, and economic well-being.

First, the services provided by delta cultural ecosystems are essential to managing natural resources. Seagrass beds, mangroves, and salt marshes are delta habitats offering abundant biodiversity and ecological services (Chakraborty 2017). Natural resource management aims to maintain their sustainability and vitality to guarantee these ecosystems' continuous provision of delta cultural ecosystem services. For instance, preservation and restoration can improve the biological functions of delta wetlands and agricultural areas, which promotes leisure opportunities and cultural diversity (Alikhani et al. 2021).

Second, the services provided by delta cultural ecosystems directly impact human well-being. By offering opportunities for leisure and recreation, they improve people's mental health and social cohesiveness. Participating in ecotourism and conservation instruction-based activities allows people to unwind, acquire new information and skills, and enhance their responsiveness and practical capabilities regarding ecological issues. Delta's cultural ecosystem services are also closely related to education and cultural practices. Figure 11.1 shows the interaction of delta ecosystems, human activities, and resources targeting the United Nations Sustainable Development Goals (United Nations 2022).

Furthermore, there is a strong correlation between economic activity and the cultural ecosystem services provided by deltas, especially in ecotourism and fishing. In addition to giving local communities financial support, these endeavors encourage the long-term growth of socioecological systems (Kokou et al. 2024). However, human endeavors like port traffic and coastal development can put delta ecosystems under stress, a phenomenon known as "delta pressure." Therefore, effective management of natural resources is necessary to balance ecological preservation and economic growth.

In conclusion, the effects and interconnectedness of natural habitat management, social well-being, and delta cultural ecosystem services form a mutually strengthening structure. Effective natural habitat management is essential to ensure the continued facility of delta cultural ecosystem services, directly improving human

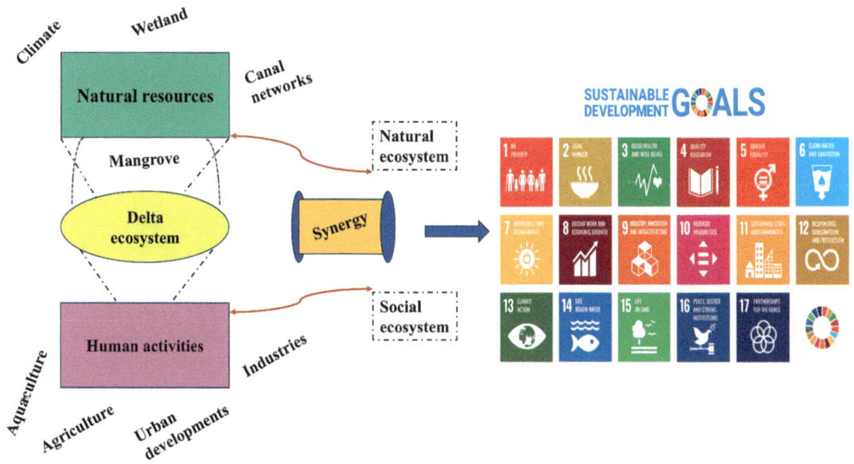

Fig. 11.1 Interaction of delta ecosystems, human activities, and resources. (United Nations 2022)

well-being. At the same time, increased human well-being encourages the prudent use and preservation of natural habitats. This synergy highlights the significance of attaining sustainable growth in delta areas.

11.3 Research Trends of Delta Cultural Ecosystem Services

It is clear from the correspondence between various delta ecosystem types and cultural, ecological services that the cultural ecological services offered by delta ecosystem types can be influenced by a variation of factors, including the needs and cultural traditions of coastal communities, environmental burdens, environmental vulnerabilities, and the structural and functional differences of various coastal ecosystem types. Studies on aquaculture and wetlands are more numerous than on other coastal ecosystems. This is because aquaculture offers human beings both financial advantages and food supplies. Furthermore, human activities like intense aquaculture farming, pollution, and climate change pose several hazards to wetlands and mangroves. One of the current study goals is to comprehend how these risks affect aquaculture and ecosystems and how to safeguard these natural resources through public participation (Frankic and Hershner 2003). There is much research pertaining to urban growth when it comes to cultural and ecological services. This is because governments and other stakeholders must think about how to successfully manage and preserve cultural assets while also striking a balance between the need to conserve ecosystems and infrastructure development (Antrop 2005). Additionally, tourism and transportation frequently draw larger crowds, raising public responsiveness and preserving ecosystems and social tradition.

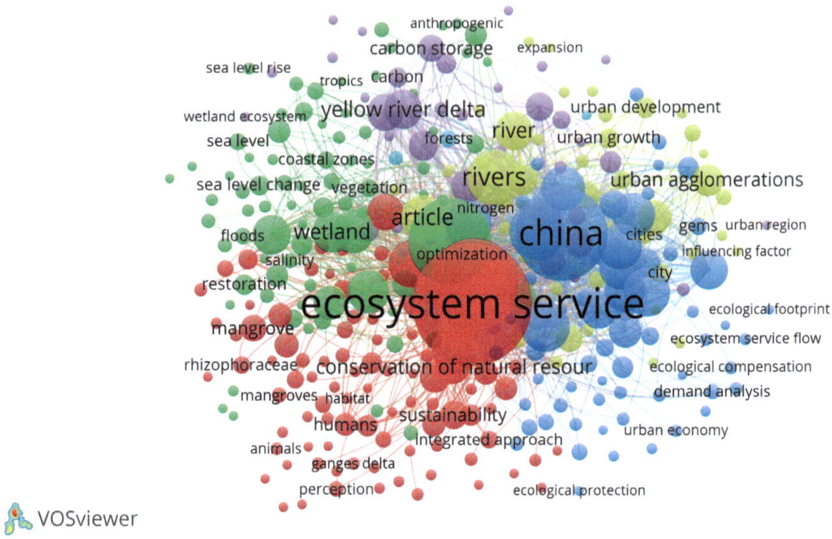

Fig. 11.2 Keyword co-occurrence representation for delta cultural ecosystem services research

Keyword mining can be used to investigate research trends and the evolutionary background in a certain field. Co-occurrence visualization processing was applied to terms with a co-occurrence frequency greater than two with the aid of a VOSviewer. Figure 11.2 clarifies this by eliminating keywords like "delta ecosystem services" associated with literature search inquiries. It affirms the importance of nature highlights the duty of humans to the environment promotes the conservation of biodiversity and social tradition backs sustainable growth and motivates positive measures to safeguard these social facilities. By examining the behavioral patterns and perceptual preferences of many society actors in a range of social circumstances research on cultural ecological services functions at both the spatial and local levels investigating efficient avenues for local resource management. Delta cultural ecological services research is an interdisciplinary field that applies economics geography ecology and sociology theories to problems in several fields.

11.4 Diversity of Delta Cultural Ecosystem Services and Policy Implementation

To completely comprehend the meaning and value of delta cultural ecosystem services, ecological, economic, and social variables must be thoroughly taken into account due to their complexity and diversity. The development of efficient communication channels and cooperation between ecology, culture, sociology, and economics is necessary for such integrated analysis to promote comprehensive studies of the cultural ecological services provided by deltas. Delta regions are particularly vulnerable to natural disasters because of their geographic site and climatic environments. These threats include coastal pressure and sea-level rise. With its information on delta ecosystem functioning, ecology offers a thorough understanding and analysis of delta ecosystems. Additionally, the contributions of distinctive ecosystems and natural habitats to local delta cultural ecosystem services are examined, along with changes in delta cultural ecosystem services in the context of climate change and sea level rise. Utilizing delta cultural ecosystem services as instruments to enhance comprehension of marine cultural heritage and highlight cultural diversity's influence on delta cultural ecosystem services, cultural studies concentrate on elements like delta culture, traditional traditions, and cultural legacy. The significance of inland aquaculture to delta communities' community and social well-being is highlighted by sociology's examination of social aspects such as community groups, societal interactions, and cultural identity. Sociology also explores shifts in human emotions. By examining the effects of economic doings like tourism and inland aquaculture on delta cultural ecosystem services and attempting to improve resource distribution to meet human society's needs, economics tackles issues about the worth of ecotourism, urban development, and delta cultural ecosystem services.

The functions and makeup of delta cultural ecosystem services are quite complicated, encompassing a variety of factors like human activity, regional distribution, and biodiversity. Ecological preservation and economic growth are frequently traded off in delta areas. The management of delta areas must consider several variables, including social well-being, economic development, and ecological preservation. Only by doing multidisciplinary research on delta cultural ecosystem services can we guarantee the thoroughness and efficacy of policy execution and offer a scientific foundation for policy formation (Peng et al. 2024). Ecology, for example, offers a basic comprehension of ecosystem architecture and functioning, but cultural trainings show how human standards and customary practices affect ecosystem facilities. Sociology examines the relationship between delta cultural ecosystem services and social structures and behavioral patterns, while economics evaluates the services' economic worth and benefits to human welfare (Duku et al. 2022). These disciplines' interactions enhance our multifaceted understanding of the cultural ecosystem services provided by deltas. The application scenarios of delta cultural ecosystem services can be expanded through interdisciplinary addition, which allows information and procedures from various academic organizations. It also offers theoretical provision and applied direction for accomplishing the dual goals of delta ecosystem balance.

11.5 Conclusions

In order to offer a fresh viewpoint on the dynamics and complexity of delta ecosystems, this research takes a methodical approach. As a result of this research, we have a more thorough grasp of the diverse benefits of delta cultural ecosystem services in preserving ecological equilibrium, fostering human well-being, and advancing economic activity. In particular, this study thoroughly evaluates existing literature based on the keyword co-occurrence trends with the relationship between delta cultural ecosystem services, human activity, and natural resources.

In order to manage delta cultural ecosystem services effectively, interdisciplinary cooperation is needed to integrate ecological, socioeconomic, and public perception data. Addressing data-gathering obstacles such as cost, representativeness, privacy concerns, and biases on social media is critical. Involving stakeholders guarantees inclusive policymaking that balances socioeconomic well-being and ecological preservation.

The functional equivalency strategy incorporates policy-based levels, resource sustainability, feedback, estimated metrics, and stakeholder engagement to successfully transition to ecosystem management. Decision-making frameworks for ecosystem management can be operationalized through the use of the functional comparability approach to maintaining delta ecosystem services.

References

Alikhani S, Nummi P, Ojala A (2021) Urban wetlands: a review on ecological and cultural values. Water 13(22):3301

Antrop M (2005) Why landscapes of the past are important for the future. Landsc Urban Plan 70(1–2):21–34

Bai X, Van Der Leeuw S, O'Brien K, Berkhout F, Biermann F, Brondizio ES et al (2016) Plausible and desirable futures in the Anthropocene: a new research agenda. Glob Environ Chang 39:351–362

Bousquet F, Le Page C (2004) Multi-agent simulations and ecosystem management: a review. Ecol Model 176(3–4):313–332

Brondizio ES, Vogt ND, Mansur AV, Anthony EJ, Costa S, Hetrick S (2016) A conceptual framework for analyzing deltas as coupled social–ecological systems: an example from the Amazon River Delta. Sustain Sci 11:591–609

Chakraborty SK (2017) Ecological services of intertidal benthic fauna and the sustenance of Coastal Wetlands along the Midnapore (East) Coast, West Bengal, India. In: Coastal wetlands: alteration and remediation, pp 777–866

Chan KM, Satterfield T, Goldstein J (2012) Rethinking ecosystem services to better address and navigate cultural values. Ecol Econ 74:8–18

Cheng X, Van Damme S, Li L, Uyttenhove P (2019) Evaluation of cultural ecosystem services: a review of methods. Ecosyst Serv 37:100925

Duku E, Mattah PAD, Angnuureng DB, Adotey J (2022) Understanding the complexities of human well-being in the context of ecosystem services within coastal Ghana. Sustain For 14(16):10111

Frankic A, Hershner C (2003) Sustainable aquaculture: developing the promise of aquaculture. Aquac Int 11:517–530

Hossain MS, Dearing JA, Rahman MM, Salehin M (2016) Recent changes in ecosystem services and human well-being in the Bangladesh coastal zone. Reg Environ Chang 16:429–443

Kokou K, Syampungani S, Chirwa PW, Makhubele L (2024) Socio-ecological Management of Multifunctional Landscapes. In: Trees in a Sub-Saharan multi-functional landscape: research, management, and policy. Springer, Cham, pp 267–304

Kuenzer C, Renaud FG (2012) Climate and environmental change in river deltas globally: expected impacts, resilience, and adaptation. In: The Mekong delta system: interdisciplinary analyses of a river delta. Springer, Dordrecht, pp 7–46

Moberg F, Rönnbäck P (2003) Ecosystem services of the tropical seascape: interactions, substitutions and restoration. Ocean Coastal Manage 46(1–2):27–46

Mondal S, Palit D (2022) Challenges in natural resource management for ecological sustainability. In: Natural resources conservation and advances for sustainability. Elsevier, pp 29–59

Nagaraju TV, Malegole SB, Chaudhary B, Ravindran G (2022) Assessment of environmental impact of aquaculture ponds in the western delta region of Andhra Pradesh. Sustain For 14(20):13035

Nicholls RJ, Hutton CW, Lázár AN, Allan A, Adger WN, Adams H et al (2016) Integrated assessment of social and environmental sustainability dynamics in the Ganges-Brahmaputra-Meghna delta, Bangladesh. Estuar Coast Shelf Sci 183:370–381

Oro D, Martínez-Abraín A (2023) Ecological non-equilibrium and biological conservation. Biol Conserv 286:110258

Peng K, Dong C, Mi J (2024) Dynamic research on the collaborative governance in urban and rural black-odorous water: a tripartite stochastic evolutionary game perspective. Systems 12(8):307

Pröbstl-Haider U (2015) Cultural ecosystem services and their effects on human health and well-being–a cross-disciplinary methodological review. J Outdoor Recreat Tour 10:1–13

Salmivaara A, Kummu M, Varis O, Keskinen M (2016) Socio-economic changes in Cambodia's unique Tonle Sap lake area: a spatial approach. Appl Spat Anal Policy 9:413–432

Sekovski I, Newton A, Dennison WC (2012) Megacities in the coastal zone: using a driver-pressure-state-impact-response framework to address complex environmental problems. Estuar Coast Shelf Sci 96:48–59

Syrbe RU, Walz U (2012) Spatial indicators for the assessment of ecosystem services: providing, benefiting and connecting areas and landscape metrics. Ecol Indic 21:80–88

United Nations (2022) Sustainable development goals communications materials. United Nations, New York. https://www.un.org/sustainabledevelopment/news/communications-material/

Wang B, Zhang Q, Cui F (2021) Scientific research on ecosystem services and human well-being: a bibliometric analysis. Ecol Indic 125:107449

Wood SL, Jones SK, Johnson JA, Brauman KA, Chaplin-Kramer R, Fremier A et al (2018) Distilling the role of ecosystem services in the sustainable development goals. Ecosyst Serv 29:70–82

Wu J (2013) Landscape sustainability science: ecosystem services and human well-being in changing landscapes. Landsc Ecol 28:999–1023

Index

T. V. Nagaraju, G. S. Bala (eds.), *Delta Ecosystems*, Environmental Contamination Remediation and Management,
https://doi.org/10.1007/978-3-031-97920-0

S

W

GPSR Compliance

The European Union's (EU) General Product Safety Regulation (GPSR) is a set of rules that requires consumer products to be safe and our obligations to ensure this.

If you have any concerns about our products, you can contact us on ProductSafety@springernature.com

In case Publisher is established outside the EU, the EU authorized representative is:

Springer Nature Customer Service Center GmbH
Europaplatz 3
69115 Heidelberg, Germany

Batch number: 08928641

Printed by Printforce, the Netherlands